From Holman Brothers to CompAir

The Story of Camborne's Engineering Heritage

Clive Carter and Peter Joseph

This book is dedicated to Clive Carter.

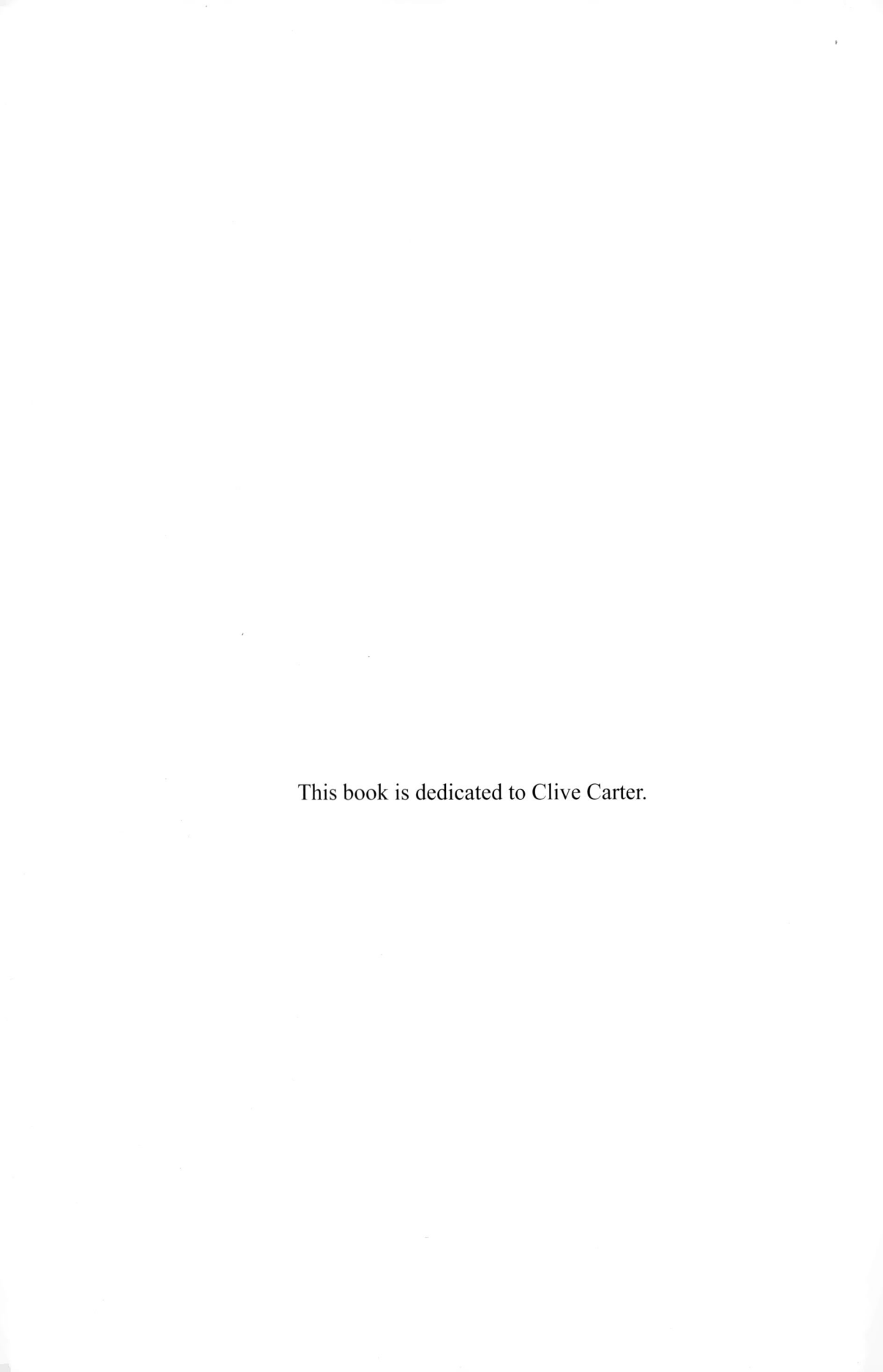

Acknowledgements

For the First edition:
N. T. Holman; A. F. M. Holman; Jim Hodge; Allen Buckley; H. Blackwell; Kenneth Brown; Eric Edmonds; Bryan Earl; Gerald Williams; David Thomas; Milton Thomas; Brian Watkins; Carol Polglase; Mrs O. P. Veal; Peter Savage; Peter Joseph, Tim Parr, Justin Brooke, T. B. Pinnington; the Poldark Mining Company; the Trevithick Society; Geevor Tin Mine Heritage Centre; the Corporation of Trinity House; the Trevithick Trust.

At CompAir UK:
Clive Tayton Operations Director; Mike Tredrea Human Resources Manager; Bill Dowse, Technical Publications Department especially Gary Tamblyn and Nigel Willoughby.

For the Second edition:
Allen Buckley; Bob Cleaves; Horace James; Brian Jones; Gavin Smith; Christine Stephens; Mike Stephens; David Thomas; Graham Thorne, Gillian Wilkins; Mansell Williams.

The Trevithick Society is very grateful to Carole Cuneo and Melanie Viner-Cuneo of the Cuneo Estate for permission to reproduce paintings and pencil vignettes by Terence Cuneo, which first appeared in Bernard Hollowood's Cornish Engineers, published by Holman Brothers in 1951.

Contents

Foreword — 10

Introduction — 11

Chapter One
A Dynasty of Engineers — 12

Chapter Two
The 'Cornish' Rock Drill — 23

Chapter Three
'Success to the Compressor' — 43

Chapter Four
The Gold Reefs — 74

Chapter Five
The Great War — 92

Chapter Six
Recovery and Revival — 108

Chapter Seven
War Again: 'The Dreaded Holman Projector' — 133

Chapter Eight
150 Years Not Out — 150

Chapter Nine
The Holman Museum   162

Chapter Ten
The Holman Sports and Social Club   168

Chapter Eleven
W. C. Stephens & Son, the Rock Drill Hospital and
the Climax Rock Drill   192

Chapter Twelve
Cornish Consolidation   235

Chapter Thirteen
Singing from the Same Song Sheet: Choirs, Orchestras and Minstrels 258

Chapter Fourteen
Broom & Wade   269

Chapter Fifteen
Reavell and Co.   281

Chapter Sixteen
International Compressed Air and World Domination   301

Chapter Seventeen
The End of an Era   313

Chapter Eighteen
Going Up Camborne Hill Coming Down   319

Bibliography   327

Index   331

Foreword

FOR over two hundred years Camborne district has been at the centre of the twin industries of mining and engineering. As mining expanded in the eighteenth century so local foundries grew to supply those mines. Blacksmiths' shops increased their range of activities and became sophisticated foundries, boiler works and engine manufacturers. At the time that Richard Trevithick Junior was working with Dolcoath's chief engineer, John Budge, William Holman was working on the mine as a blacksmith. William became father to Nicholas who is credited with founding the firm of Holman Brothers. As Perran Foundry and Harveys at Hayle produced their great steam engines, Nicholas Holman produced engine boilers at Pool, and worked with Trevithick on the high pressure boilers necessary for his engines.

From these humble beginnings Holmans grew to a position of dominance in Cornish engineering. With the closure of the mighty Dolcoath Mine, in 1920, Holmans took over as the principal employer and creator of prosperity in Camborne. Throughout the first half of the twentieth century the Camborne factories expanded and the name Holman and Camborne were recognised in every mining district on earth. By the 1960s, having taken over such firms at R Stephens & Son (Climax), the firm employed over three thousand workers - among them Clive Carter and yours truly.

The story of the first two hundred years of the firm has already been beautifully told by Clive Carter in a previous book, but this present tome takes the story right to the end. The book is three times as long as Clive's, and there are ten extra chapters to cover the many aspects of Holmans' involvement in the cultural, sporting and social life of Camborne. A vast number previously unpublished photographs have also been used.

The involvement with Broomwade and Reavell is told, together with the stories of these firms. There is a section on Holmans world famous museum, one on Blaythorne House, on the Holman-Climax Choir, and on the firm's cricket, football and rugby teams. Not just the directors but also the workers are named and described, so that long-serving families as the Godolphins and the Trebilcocks are dealt with. Indeed, almost every family in Camborne had members working at 'the foundry' in some capacity. Peter Joseph has produced a book which does Holmans and Camborne proud. Clive Carter would have been delighted with the result.

Allen Buckley

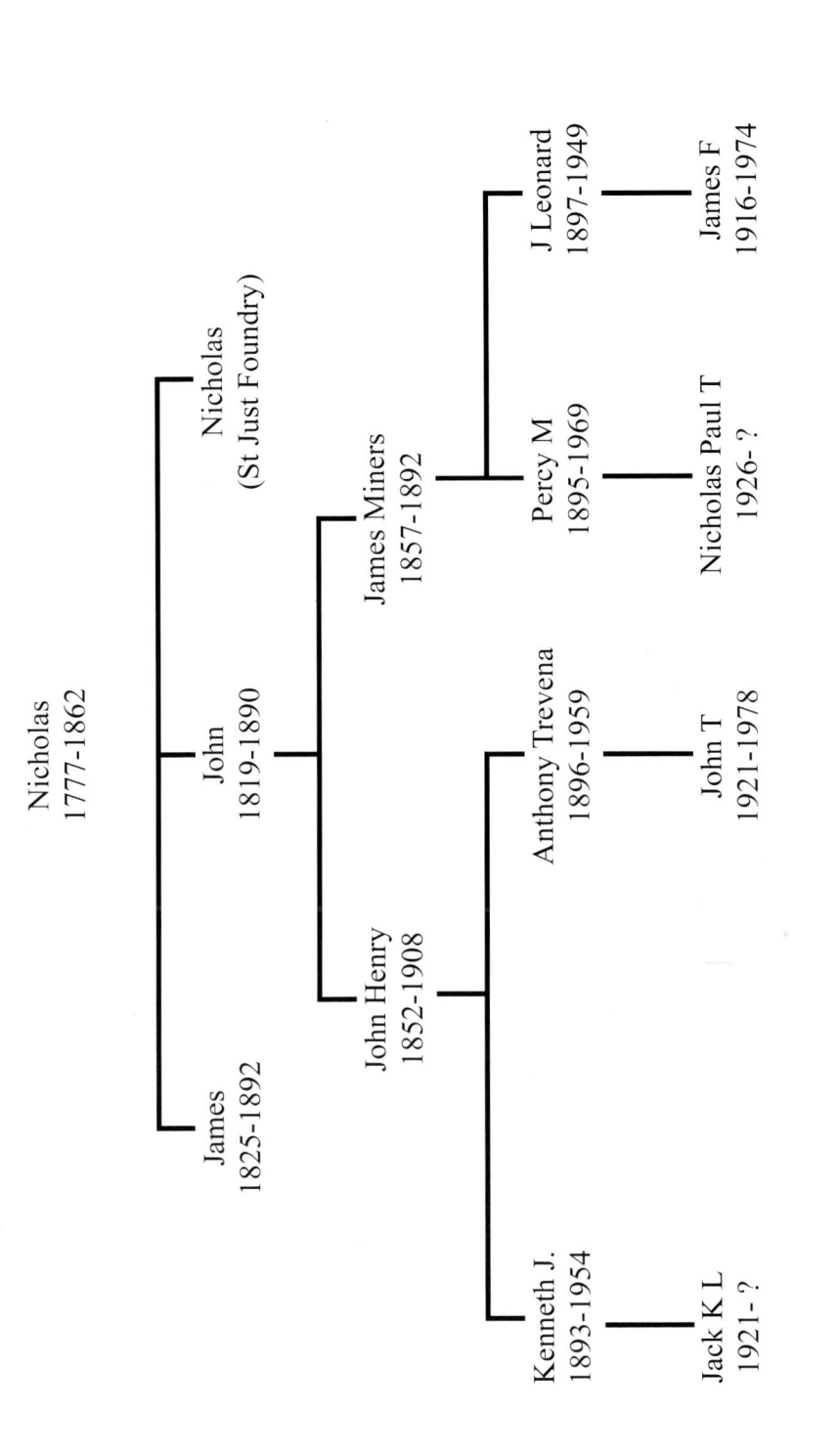

```
Nicholas
1777-1862
├── James
│   1825-1892
└── John                                    ── Nicholas
    1819-1890                                  (St Just Foundry)
    ├── John Henry
    │   1852-1908
    │   ├── Kenneth J.
    │   │   1893-1954
    │   │   └── Jack K L
    │   │       1921- ?
    │   └── Anthony Trevena
    │       1896-1959
    │       └── John T
    │           1921-1978
    └── James Miners
        1857-1892
        └── Percy M
            1895-1969
            ├── J Leonard
            │   1897-1949
            │   └── James F
            │       1916-1974
            └── Nicholas Paul T
                1926- ?
```

Holman family tree showing the principal members of the company.

First Edition 2001 © Clive Carter & CompAir UK
Republished 2001 and 2004 by the Trevithick Society

Second Edition 2012© Peter Joseph & The Trevithick Society

Printed by
Short Run Press
25 Bittern Road Exeter, Devon EX2 7LW

Layout, design and typesetting by
Peninsular Projects
PO Box 62, Camborne TR14 7Zn

ISBN 978 0 904040 94 4

All rights reserved. No parts of this publication may be reproduced, stored in a retrieval system or transmitted in any form or by any means, electronic, mechanical, photocopying, recording or otherwise, without the written permission of the publisher

INTRODUCTION

Clive Carter's book *Cornish Engineering 1801-2001* came out in 2001 to celebrate the 200th anniversary of Holman Brothers and CompAir. It was a couple of years after this that the Trevithick Society, or, more specifically myself, as Curator, came into the possession of more and more photos (among many, many, other items) concerning Holman Brothers. Climax and CompAir. I mentioned to Clive that this would be a good opportunity for someone to write a *really* good history of the companies…. those who knew Clive will realise that his reply was both artistic and unprintable!

Holman material continued to accrue to the Trevithick Society, while over the winter of 2005-2006 I recorded the CompAir buildings at the Foundry Road Industrial Estate before, and sometimes during, their demolition. During this period I also carried out some research into the latter years of CompAir and its sale to different companies prior to closure and the removal of production to Germany. Much of this appeared in an archaeological and historical report on the works, the work carried out because it was not important enough to have been carried out by the authorities.

Sadly, Clive died unexpectedly on 27th October 2006, leaving a great gap in Cornish culture and industrial heritage, as well as the Trevithick Society. It fell to me to produce the final version of *Cornish Engineering*, though with a different title (Holmans' own title, *Cornish Engineering,* was itself based on the 150th anniversary book, *Cornish Engineers*, by Bernard Hollowood). Apart from bringing the story of CompAir in Cornwall up to date, this new edition has been restyled and expanded to include sections on the old rival Climax, the various choirs and the Sports and Social Club.

I hope Clive would have been pleased with the result, though my vocabulary is extensive enough to have understood any negative comments he may have given.

PETE JOSEPH

Dolcoath Mine at the beginning of the 19th century.

Chapter One

A Dynasty of Engineers

In a large and lofty building, supported by pillars of iron, with great black apertures in the upper walls, open to the external air; echoing to the roof with the beating of hammers and the roar of furnaces, mingled with the hissing of red hot metal plunged into water, and a hundred strange and unearthly noises never heard elsewhere; in this gloomy place, moving like demons among the flame and smoke, dimly and fitfully seen, flushed and tormented by the burning fires, and wielding great weapons, a faulty blow from any one of which must have crushed some workman's skull, a number of men laboured like giants.... others drew forth, with clashing, noise, upon the ground, great sheets of glowing steel, emitting insupportable heat, and a dull deep light....
Charles Dickens *The Old Curiosity Shop* 1841

The founder of the Holman dynasty was, surprisingly, not an engineer but a man of the soil and the plough. Nicholas Holman was born at St Allen near Truro in 1716 and as a young man was brought to Tehidy Manor near Camborne to be a farm servant to the aristocratic Basset family. On Boxing Day 1743 he married Anne Rowe of Camborne and they set up home on a small farm belonging to the Bassets at Menadarva, a tiny hamlet tucked into a deep river valley sheltered from the westerly gales which drove across the bleak Atlantic cliffs of North Cornwall. The family's life must have been hard and isolated, governed by the seasons and the farming calendar. Each day Nicholas Holman would have walked to Tehidy along a rough track that ran across the moors or along the banks of the Red River. A generation earlier the river had flowed clear and fresh while trout could be caught from the ancient Menadarva Bridge. Now it was coloured red by the slimes from the Camborne mines, and the tin streams whose waterwheels and machinery crowded the banks.

This ever-present reminder of the money to be made from mining never tempted Nicholas Holman, who remained a faithful retainer of the Basset family all his life. He was a capable man who later rose to become a bailiff under Lord De Dunstanville and was also remarkably enlightened, allowing each of his eight children the opportunity to follow the career of their own choice. Emanuel remained at home and ran the farm 'as long as he was able'; Richard became a miller and Nicholas a customsman, who later had the misfortune to be drowned at sea. John and James became copper miners, his daughter Mary married a Camborne goldsmith and Grace was a tailoress who used to brush the plush and velvets of Lord De Dunstanville.

William Holman chose to become a mine blacksmith and through him arose what was to become two dynasties of Cornish engineers. He could not have chosen better, as these were the amazing years when the burgeoning Industrial Revolution forever changed the people and landscape of West Cornwall. This hothouse atmosphere of technological development created a breed of men whose sons and grandsons would make Cornish engineering famous from the Rocky Mountains to the Witwatersrand.

Mine smiths not merely shoed horses but were expected to make up and sharpen tools, work and rivet copper, brass and wrought iron and, increasingly, to assist the 'viewers'. These were travelling erectors who were setting up the first steam engines on the copper mines and demanding unheard of standards of metalworking. Methods were still crude; the Cornish engineer Robert Mitchell wrote years later that 'our blacksmiths knew nothing, comparatively speaking, of the proper method of working iron and steel, of making screw taps and plates for cutting these large adjusting screws so very useful and necessary in mining, yet so little thought on by the Cornish as by the inhabitants of the Fellow islands...' They were soon to learn, and the skill and enterprise of Cornish engineers was to outshine that of Boulton and Watt.

In October 1772 William Holman married Mary Reed and they went to live at Pool near Camborne, deep in the heart of the copper country. Besides general mining work he began to build boilers, very successfully despite the fact that it had taken decades to produce boilers adequate for even the low pressures of the early atmospheric engines. He soon had a large family, including Nicholas Holman who was born on the 3rd of August 1777; only five years younger than Richard Trevithick who lived just up the road and would become a close friend in later life. Nicholas spent his formative years working for his father and in 1801 when only twenty-four, he set up his own works at Pool to build boilers and rapidly established a first class reputation for workmanship. He may have made, or at least helped to assemble, the boiler for Trevithick's road locomotive, the famous 'Puffing Devil' which made the trip up Camborne Hill on Christmas Eve that same year. He married Ann Harvey of Illogan (whose father Henry also has a claim for helping to build the Puffing Devil)

Nicholas Holman junior, 1777-1862, founder of the engineering dynasty.

and in 1803 Nicholas Holman junior was born, one of three sons. At an early age he was taken away from Pool School to serve as yard boy and foundry messenger. He may well have had to stand before the giant figure of Cap'n Dick Trevithick himself and repeat complicated orders and receive instructions, already knowing that the engineer's kindly ways were matched by a notoriously short temper.

Nicholas senior's friendship with Trevithick brought dividends, though the great engineer's erratic business methods drove all around him to despair. In 1813 he built two of Trevithick's revolutionary Cornish boilers for a 24-inch high pressure steam pole engine which Captain Samuel Grose was erecting on Beer Alston Mine on the Tamar. Trevithick as usual fumbled the cash flow and in a letter urgently requested that 'the workmen making your boilers want an advance of cash to enable them to finish. They provide the iron and the labour for which they are to receive 42*l* per ton for the boiler when finished; the weight will be about 8 tons. You may send this money to Mr Sims or to me, or otherwise you might direct it to Mr N. Holman, boiler maker, Pool....100*l* will satisfy them for the present...' Trevithick again wrote later 'Gentlemen you have the bill for sundry expenses paid on the Beerlalston Boiler which I have advanced to the Smith, N. Holman, long since to enable them to finish them, would you wish a bill to be drawn on you by Holman for their balance due or shall I draw the bill on you and send you Holman's receipt? ...'

Trevithick came to Nicholas Holman for boilers for the high-pressure engines which he contracted to ship out to the Peruvian silver mines. On 22nd May 1813 he wrote 'A great part of the wrought - iron work and the boilers I have arranged in Cornwall'. Just a few weeks later in typical whirlwind fashion, Trevithick writes again that 'Yesterday I engaged all the boiler plate in the county, which will be sent today to the different workmen. The master smiths that I have engaged are the best in the kingdom. I have obligated them to put the best quality of iron, and to be delivered at Falmouth within four months. I have been obliged to give them a greater price than I expected, otherwise they would not turn aside their usual business for a short job of four months'. Obviously Nicholas Holman, despite his respect and friendship for Trevithick was becoming very wary of the engineer's disregard for money and the former soon made arrangements that 'boiler builders here, to weigh and pay at the end of every week'.

The plates were not to be assembled as, like the engines and the rest of the machinery, they had to be carried up the mountain tracks of the Andes on the backs of mules, plate by plate, and riveted up on the mines. The boilers were ready by the end of September 1813 and formed part of a very large consignment, recorded in Trevithick's account book as being shipped from Penryn.

Holmans business continued to prosper and he took a contract to build boilers for Wheal Vor near Helston. Young Nicholas was sent across to prepare the tools and look after the money as Thomas Jenkins, the Pool foreman, was illiterate. Soon afterwards Nicholas senior was approached by Robert Mitchell, one of the proprietors of the new Copperhouse Foundry at Hayle, already a rival to the Harvey Foundry just down river. He was offered the job of manager of the boiler works but, now middle-aged, he was not inclined to leave his home and

James Holman junior, 1825-1892, youngest son of Nicholas.

business at Pool. Instead he suggested that young Nicholas, subject to his father's help, took the position. Nicholas senior thoroughly trusted his son and rarely visited Copperhouse and when he did, he 'took up his portion with clean hands never having been called upon to do a days work.

When Nicholas junior left Copperhouse to set up his own foundry at St Just in 1834, he 'recommended my brother William as a proper person to succeed me and offered myself to render him assistance at any time if required, which was agreed to'. William was then only twenty-one but such was the training he received at their father's works that he was readily taken on at Copperhouse.

James Holman, the youngest brother, born in 1825, would eventually manage the Pool works for his father and would be the only one to remain a bachelor. John Holman, the middle son, born in 1819 was scarcely twenty when, doubtless enthused by his brother's success at St Just, he decided to set up his own business. In 1839 he established a small boiler works and foundry at Camborne and this eastern end of the town would for the next one hundred and seventy years become in effect 'Fort Holman' with all the noise, bustle and

John Holman, 1819-1890, founder of the Camborne works.

John Henry Holman, 1852-1908.

smell emitted by a busy foundry.

Camborne had changed considerably since the days of Trevithick. From a few houses lining a muddy street around the church it had grown into a busy small town that would soon be on the new railway line up from Hayle to Redruth. Indeed, John Holman's foundry was built beside the magnificent Wesleyan Centenary Chapel that gave its name to the new street. He was a very shrewd and excellent businessman, whose success paralleled that of his brother's foundry at St Just, where the legend 'If it's metal take it to Holmans' might easily have been coined for both small foundries. Unlike his brother, he was far better at mining speculation and invested in most of the Camborne mines, he enjoying the privilege, like many other entrepreneurs, that as both supplier and shareholder he rarely lost money.

He continued the family speciality of building the iron plate boilers which generated the high-pressure steam needed by the pumping engines of the numerous mines between Carn Brea and Lands End. Yet neither he nor Nicholas saw themselves as rivals to the three great Cornish foundries of Perran, Harvey and Copperhouse, where the engineering heirs of Woolf and Trevithick brought the Cornish high-pressure steam engine

James Miners Holman, 1857-1953.

to perfection and worldwide use. In contrast, although they were both heavily involved in making boilers and mining machinery, the Holmans relied upon a steady and well-founded, if unspectacular trade which embraced a vast variety of other customers. John Holman made agricultural implements like his cousin, who had built the first iron plough in West Cornwall at his new wharf-side Foundry in 1841, as an advert in the *West Briton* of 1846 demonstrates.

> 'John Holman. Boiler Builder and Farming Implements manufacturers, Camborne, begs most respectfully to call the attention of his friends and agriculturists generally, to his improved portable and other Potato and Turnip steaming apparatus, which for cheapness, utility and economy he is confident cannot be excelled'.

Old Nicholas Holman died in August 1862, leaving the Pool Works to his brother James, but by then John Holman's own foundry was well established. His first son, John Henry, was born in 1853 and the second, James Miners, in 1857. Both were destined to make the name 'Holman Brothers' famous in every mining camp in

John Holman's works at Camborne in the early 1840s.

The Camborne works in the late 1860s or early 1870s. The Centenary Chapel is on the right. Compare with illustration opposite.

the world. They duly did their apprenticeships; James doing his at Camborne but John went to the famous Birmingham 'Cornwall' Works of the expatriate Cornish engineers, the Tangye family. George Tangye, one of the partners, actually did his own apprenticeship at Holmans before he and his brother James migrated to the Midlands to establish their own engineering empire.

By the time the Holman boys finished their apprenticeships, the fabled Cornish 'Copper Kingdom' was fast sinking into ruin as the copper mines, already forced to sink deeper in pursuit of their waning mineral riches, were rivalled by vast copper strikes in Australia and on the Great Lakes of North America. To worsen matters, the price of tin began the often-violent fluctuations which almost destroyed Cornish mining by the turn of the century. Depression and slump made inevitable the fate of the big Cornish foundries, whose fortunes were far too closely wedded to those of the ailing copper mines. There were the first warnings that the era of the great 'Cornish' pumping engine was ending, as these once innovative and technological wonders were left bobbing in the wake of the ship of progress, which would increasingly rely on oil, electricity and compressed air. Perran and Copperhouse Foundries had both gone by the early 1870s, while the Harveys of Hayle embarked on shipbuilding to fill the gaps in their order books, but the glory days of their massive foundry were almost over.

In 1870 the Holman sons, despite their relatively young age, had joined John as directors of their company. They were completely complementary to each other; John Henry, brilliant and innovative engineer and James Miners, a very shrewd and astute businessman. Together they laid the foundations of what would become a worldwide company and the tool of their success was the rock drill, or 'boring machine' as it was then popularly known.

The Camborne Works of Holman Brothers, circa 1887. Centenary Chapel again on the right.

John Holman remained as versatile and resilient as ever and was easily able to survive these hard times and his price list for 1880 demonstrates that Camborne Foundry & Engine Works could produce engine cylinders and beams, stamp gearing and flywheels, stamp heads, wagons, kibbles, skips, and picks and shovels; while 'Plain Cylindrical Boilers, made of the Best Plates and Best Rivet iron' were 18/- per cwt.

To transport their foundry material and other goods, John Holman bought a traction engine, which in January 1885 hauled 13 tons of tin concentrate to the smelter at Point, near Devoran. This was the first load to arrive behind a traction engine, and the trip was accomplished despite heavy frost on Brea Hill and around Tuckingmill. Often the engine was sent down to their Holman cousins at St. Just for boiler repairs. John Holman, who always prefaced his letters to Camborne 'Dear Cousins', unashamedly used his influence as a shareholder in the St Just mines to win contracts for them as in January 1887 when he wrote regarding a big job for Botallack, hoping that 'the Lion's share of it will come to our beloved cousins in Camborne'.

The new traction engine made itself better known just two years later in Truro. The engine, pulling two trucks carrying one side of an engine beam destined for a mine to the east of Truro, ran away down an incline between Boscawen Street and Nicholas Street. Damaging the kerb, the trucks only stopped a foot short of the chemist, Mr Vincent, after demolishing the lamp post in front of Clarke & Co. The movement of the trucks pushed the engine out so that it blocked Nicholas Street, 'which made the task of uncoupling very difficult. It was eventually accomplished without the breakage of any gear, and after some further manipulation, in which

This photograph from 1887 apparently shows the entire workforce of the Works. The figure to the left with the enormous beard is the Company Secretary.

the engine was dexterously managed by the driver, the trucks were again placed in position, and the iron horse, with its heavy burden, proceeded on its journey.'

In 1894, the young George Bertram Tangye joined the company, having spent a short time at the Grenville United Mines. After serving his apprenticeship in pattern making he would work as assistant to the legendary Nicholas Curry before taking charge of the Smiths' Shop and Shipping Department before retiring in 1947. Born at Gwinear in 1870, 'Mr Nicky' joined Holmans at the age of sixteen and rose from a humble apprentice pattern maker to become foreman of the Pattern Shop in five years.

With the exception of the Dorothea engine (see below) John Holman never built really big Cornish engines, though in 1897 a 28-inch engine was built for Dolcoath stamps. Usually he supplied small handy engines that attracted a wide variety of customers from Tehidy Saw Mills to Trevarno tin streams. John Mills, a Camborne builder, praised his Holman engine which 'drove a 30-inch saw with perfect ease and the amount of work that can be done with proper appliances is astounding'. Even Michael Loam wrote that the 'little horizontal engine you sent us this summer is working exceedingly well, and shall be happy to recommend it'.

A tattered notebook provides an insight into the everyday work of the foundry;

> 21st Dec. Make one Bevel Wheel East Pool Pulveriser at 4 days at 1/6- Commenced Monday ½ day.
>
> W. F. J. Harvey make core boxes and sweep for centrepiece Dolcoath Stamps. Also alter arm. ½ day'.
>
> 9th Nov. John Veale had 6 days to make 3 foot 6 Slab with lining best brass work etc. for Capt. Rabling
>
> W. H. Carbis make 12 Bell mouth kibbles at 6 days 2/6- Carbis 1/6- Boy. commenced Friday 10 am. Gained 2/6-. boy 1/6-. ½ day.
>
> N. Curry make 12" Roll 3" wide. Time 4 ½days. Commenced 1 Aug. 1 30 pm.

Drilling an 'upper' by hand, or 'beating the boryer'.

Chapter Two

The 'Cornish' Rock Drill

The true percussive rock drill was essentially an American invention, born of the vital need to break a technological impasse that threatened the great designs to span that continent. Railroad engineers and contractors had to pierce hills and mountains on a scale and speed never before envisaged and could not afford to drive their tunnels by hand labour. In the words of the famous American ballad, *John Henry** might have been determined 'not to let that steam drill take my place', and to 'die with a hammer in my hand', but the laborious method of drilling by hand to tunnel through solid rock was about to be revolutionised.

Fifty years before, Richard Trevithick had invented a rail-mounted steam-driven rotary borer which helped quarry large blocks of limestone for the Plymouth Breakwater. In 1838, the American Singer brothers had built a machine where the drill was raised by a steam piston and allowed to drop by gravity, boring what was known in Cornwall as 'a downer'. Singer drills were used on the Illinois and Michigan Canal and in enlarging the Erie Canal. Although the brothers never pursued their ideas, Isaac went on to perfect the sewing machine.

The real ancestor of the modern rock drill was the Lance, patented on 27 March 1849 by John Couch, surely a Cornish expatriate, of Philadelphia. His steam driven machine drill used a heavy flywheel, ratchet and a piston to create a momentum which drove the drill 'lance like' against the rock face at any angle. On the rebound it was caught by a gripper and the forward stroke of the piston

***John Henry was a legendary steel driving man, whose life is the basis for one of the world's best-known folk tales. His fame rests on a single epic moment when he raced a steam drill during the building of a West Virginia railroad tunnel.**

Jordan's hand-powered rock drill of 1871.

hurled it forward again. Two years later J. J. Fowle, his principal assistant, took the design a stage further and became the first engineer to propel the drill against the rock by the direct action of steam upon a piston, to which the drill was attached, a motion that would feature in rock drills for the next fifty years. He also used a ratchet and pawl to rotate the drill steel and replaced the Lance motion by setting the drill on a cross head, making it far steadier. The attachment of the drill steel directly to the piston was to be a standard feature of rock drills for decades.

Fowle's basic designs were developed by other engineers like Charles Burleigh, who eventually bought Fowle's patent to avoid potential infringements. The Burleigh drill weighed in at 240lbs and delivered 200 blows per minute. Forty of his big steam drills were used to drive the Hoosac Railroad Tunnel under the hills of western Massachusetts by General Herman Haupt, a veteran US Army Engineer, who had built roads through the wilderness and had been promoted for services to the Union after the Battle of Bull Run. Despite the size and weight of these machines the 40 units had to be repaired 250 times a month! He preferred compressed air to steam, and the appalling underground conditions were then so radically improved 'that everybody was convinced that air was preferable and no more talk of steam was heard'.

In 1870, Simon Ingersoll, formerly a farmer, accepted a contract to design a drill able to work on rock. He based his work out of a New York machine shop owned by Jose F. de Navarro. By the following year, Ingersoll had developed and patented a rock drill; however, it proved not to be sufficiently strong for New York's streets. The same year Ingersoll

The Burleigh rock drill of 1871.

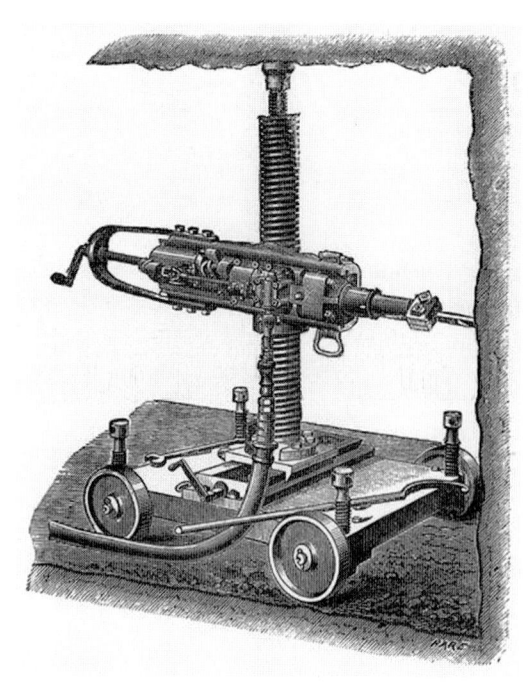

Burleigh wagon mounted rock drill from 1872.

acquired the Fowle-Burleigh patents and the Ingersoll company merged with that of Burleigh. Henry Clark Sergeant, a partner in de Navarro's machine shop, saw an opportunity to improve upon Ingersoll's drill, making it more resistant to breakage by separating it into two components. Following this enhancement, Sergeant convinced de Navarro to purchase Ingersoll's patent. This event led to the 1874 development of the Ingersoll Rock Drill Co. with Sergeant acting as the company's president. Ingersoll left the company around this time and the credit for producing the most famous rock drill ever went to Harry Sergeant. Sergeant's company made a drill which used the revolutionary Eclipse valve, which had been borrowed from a pump manufactured by one of his neighbours. In 1884 the two companies merged to form the famous Ingersoll-Sergeant Rock Drill Company. In 1905 the Ingersoll-Sergeant Drill Company merged with the Rand Drill Company to form Ingersoll Rand, which still operates.

In the meantime, Simon Ingersoll, though responsible for the drill's initial development, continued to live in virtual poverty, relying on the sale of his remaining patents for the survival of his family until his death in 1894. As a tribute to Ingersoll's leadership and technological contributions, rather late in the day, he was elected into the Mining Hall of Fame in 1992.

The Ingersoll 'Eclipse' rock drill, 1874.

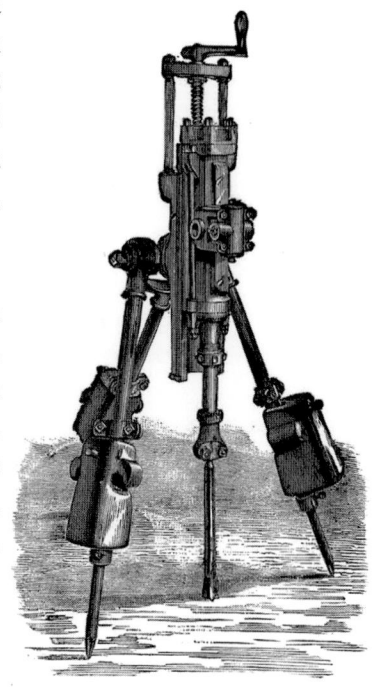

Someiller's rock drill in use in the Mont Cernis tunnel.

A greater epic than the Hoosac was the driving of the eight-mile tunnel to take the railway from Paris to Turin under the 12,000 ft. Mont Cenis in the Italian Alps. Work began in 1857 and hand labour and gunpowder blasting produced a weekly advance and casualty rate reminiscent of a bad spell in the siege lines outside Sevastopol. Germain Someiller, the engineer commanding the French heading at Fourneaux in Savoy, produced his own large machine drill, with air being supplied by compressors of his own design, driven by large waterwheels. There were many setbacks and the work was very slow by modern standards, but the Mont Cenis Tunnel was finally driven through on Christmas Day 1870 by Borer No. 45, a massive multi-drill rig mounted on a railway bogie and towing its own water tank to de-sludge the holes.

All this innovation and expertise was displayed to great acclaim at the Paris Exhibition of 1867 and created extraordinary interest among the more forward thinking Cornish mining men. To many, this sudden outburst of mechanisation was the answer to the desperate state of Cornish mining. 'Cousin Jack'

The 'Victor' rock drill of 1879.

26

THE BARROW ROCK DRILL

As worked at Dolcoath Mine in the hardest rock in the world.

PROCEEDINGS OF THE MINING INSTITUTE OF CORNWALL.

An early advert for the Barrow rock drill.

was going to suffer the same fate as 'John Henry'. Later that same year, many of those who exhibited at Paris came to Cornwall with their machines, including General Haupt. The Cornishmen seemed to have been mildly bewildered by the plethora of machine drills offered to them by a variety of entrepreneurs, all eager to test their machines in such a prestigious mining country. Many more conservative Cornish mining men opposed mechanisation, but in 1868 Captain William Teague, a legend in Cornish mining, invited Frederick Doering of Rhurort in the Ruhr to test his machine at Tincroft Mine. Despite initial and spectacular success the Doering drill was a failure, not only there but also at Dolcoath, where the formidable Captain Josiah Thomas had arranged a trial for the German machine.

There was a rush of improvisation to the rock drill not dissimilar to that to the Cornish beam engine after the end of Watt's patent in 1800, and between 1850

The Barrow rock drill.

and 1875 over 110 rock drill patents were taken out in the United States and 86 in Europe; 64 of these were in Britain.

Many mining men became convinced that a special drill was needed for the Cornish mines and their quest for such a machine brought many drill companies down to the Cornish mining country, especially when another slump in tin made mechanisation even more urgent. Captain Josiah Thomas of Dolcoath Mine, a staunch champion of the rock drill, chose the Barrow, developed by James Hosking, 'a smith we sent to Barrow about twenty years since' and Harold Blackwell, an engine smith who also worked at the Park Iron Pit, near Barrow-in-Furness. With a heavy cast gunmetal body the Barrow, which was cranked forward along the cradle by the same handle that rotated the drill steel, began driving the 314 level west in Old Sump Shaft in October 1875. The drill had to be

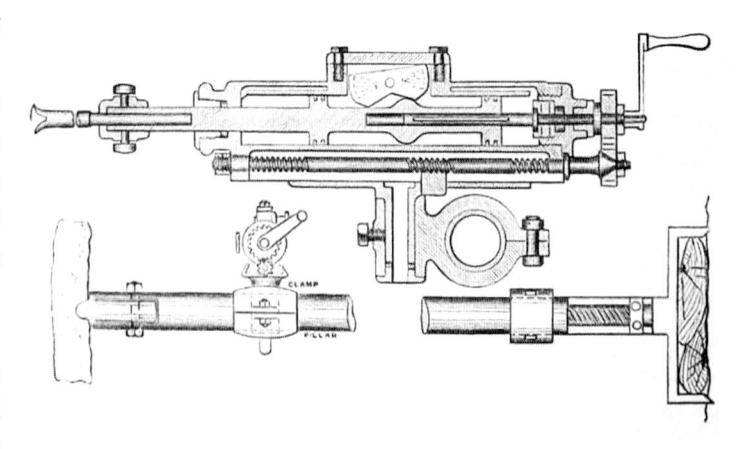

Wagon-mounted Doering rock drill.

Holman Brothers' tappet rock drill, 1881, photographed outside No.3 Works.

redesigned after receiving a fearsome pounding, boring rock so hard 'it would scratch like glass'. Yet the Barrow crew drove four times faster than the hand labour team driving a level 12 fathoms above their heads. Captain Thomas was delighted and bitterly criticised the die-hards who opposed the rock drill, likening them to those who preferred to cross the Atlantic in a sailing ship while he preferred sailing on 'a Cunarder'. He even allowed Leila Noble of Rome, Georgia, granddaughter of a Cornish emigrant, to inspect the Barrow at the bottom of New Sump Shaft when she went underground down the man engine and 30 fathoms of ladder roads at Dolcoath in June 1879.

The first rock drill to be designed and built in Cornwall was the Climax, the brain-child of Richard and William Charles Stephens. The two had set up the 'Rock Drill Hospital' at Pool to repair the drills made by other manufacturers. The company of R. Stephens & Son was formed on 1st October 1878, the Climax rock drill appearing the following year. There was to be an intense rivalry between Climax and the Holmans, which would only dissipate after the Climax company was taken over nearly 70 years later.

More rock drills were offered to the Cornishmen who must have keenly read their makers' superlative adverts in the *Mining Journal*, or the letters that reflected the bitter rivalry between the companies themselves. The American McKean, a development of the General Haupt tunnelling drill, was put to work at South

Very early type of diamond drill.

Roskear and another American drill, the Burleigh, in Rocks and Goonbarrow Mine. Carn Brea chose the enormously powerful rock drills of the celebrated Major Beaumont R. E., of the Diamond Rock Boring Co. who specialised in tunnelling and deep boring. The cost was appalling and the Beaumonts were really not suitable for Cornish mines. Its only distinction was to be the first machine to be responsible for a fatal rock drill accident when a miner drilled into a miss-fired shot in September 1879. Captain Thomas had already contracted with another rock drill company, Ulathorne & Co., to use their Champion drill, and firmly believed 'the time is not far distant when people will not think of working Cornish mines without boring machines'.

This technological revolution was not lost on the Holman boys, but how they became converts to machine drilling is a matter of conjecture; most likely they saw a gap in the market which they believed they could fill. They were both astute Cornishmen who realised that mechanisation and mining were inseparable, and that the future meant radical change, if they were not to share the fate of the big foundries. Their Camborne works was far less vulnerable but the demand for mining machinery in Cornwall itself was shrinking and the Holman boys became convinced that compressed air, rather than steam, was the key to a worldwide market. They already made very successful air compressors and it must have seemed quite logical to make rock drills as well. Such was their enthusiasm that their father became so heartily tired of their talk of rock drills that he forbade them to speak of them over the breakfast table.

Another Burleigh rock drill, this one from 1885.

John Holman retired in 1880 and the boys took over the Camborne works and immediately began experimenting with rock drills. Already in Cornwall was James McCulloch, an English-born Scottish mining engineer, who around 1870 had worked for the Burleigh Rock Drill Company. He had also spent several years as a contractor driving levels at the Rio Tinto in Spain. Abrasive and self-opinionated, McCulloch's expertise with rock drills was undeniable and he designed several machines and tunnelling carriages. He had recently arrived as an agent for Hathorn & Co., whose very successful Eclipse rock drills had won a Silver Medal by drilling four times the speed of hand labour in a contest held at Falmouth Docks in August 1879. The Cornishmen had reservations, fearing the drills' valves were too sensitive and they also had their pet hate, automatic feed. Yet within a few months the Eclipse had captured East Pool Mine, second only to Dolcoath, driving the 180 west to uncover a huge run of tin ground. The Eclipse even penetrated the conservative mining country around St Just, when the Botallack adventurers chose it to drive northwards into Wheal Cock, which was being developed after the abandonment of the famous Crowns section.

Testing a rock drill at the Camborne works in 1881.

In the late 1880s another major advance in rock drills occurred when C. H. Shaw, an engineer in Denver, Colorado, devised a hammer drill in which the piston was separated from the drill rod, which it hammered in each cycle. This increased the frequency of the blows and thereby the rate of penetration. While this was suitable for holes drilled upwards, because the cuttings fell out of the hole by gravity, it did not help with downward holes, which clogged with mud and grit.

The first Holman-McCulloch rock drill: McCulloch, centre, is flanked by James Miners and John Henry.

This problem was not solved until 1897, by J. G. Leyner. George Leyner was a Colorado farm boy turned engineer who undertook considerable experimentation on rock drills. Unfortunately his first attempt was a complete disaster. The first drill utilised a hollow drill steel through which air was blown to remove the rock cuttings. The removal of the rock fragments was the last major development towards the modern rock drill. The patent for the drill was granted in 1897 and he sold 75 machines straight away. These almost ruined him as miners refused to work with them, the amount of air and chippings being blown through the steel making work impossible. Back to the drawing board, Leyner came up with the idea of a 'water tube', a small diameter steel tube designed to pass down through the centre line of all the interior components - like the piston - until it located in a similar hole in the centre of the drill steel itself. Water pumped down the tube turned the dust into a sludge that trickled out of the borehole.

How McCulloch met the Holman brothers is unknown. The Eclipse agents were often at Dolcoath where John Holman was a shareholder; in May 1879 he was arguing with Captain Thomas regarding the cost of replacing the underground tram iron rails with steel. There was also a big meeting at Abraham's Hotel at Camborne, which was attended by all the mine captains and engineers including Captain Henderson of the Eclipse.

McCulloch may have already abandoned the Eclipse and joined the Holmans, as in October 1880 they filed a patent for an 'improved machine for drilling rocks'. Building the prototype was no easy task as, although the Holman foundry was well equipped, a rock drill required new jigs and tooling, and the tolerances required for compressed air were finer than for steam pressure. There were patent drawings but the standard procedure was to make the part first, then the drawing and the pattern.

In April 1881 the Holmans signed their first agreement with James McCulloch. The witness, interestingly, was the first works manager, Alfred Harvey, who had been apprenticed to their father in 1847 at twelve years of age and three years later had begun his foundry career in the smiths' shop. The agreement stated that, once manufacturing costs were deducted, the actual selling price would be then mutually determined and the profits would be shared equally.

James McCulloch posing with the model of the new rock drill. The smile on his face (or is it a smirk?) was not to last!

The Holmans would advance sufficient money to McCulloch to secure the patent on his tool holder and also finance other developments, but the cost of marketing and publicity would be shared equally. The new Cornish rock drill made its debut at a rock drill contest held at Dolcoath Mine by the Mining Institute of Cornwall on 8th December 1881. Drilling against the McCulloch's old partners, Eclipse, and a Barrow brought up from underground, the 'Cornish made a winning debut', and 'by greater durability and special adaptability to be more suitable for the mines of the West than those now in use, a point which experience, of course, can alone satisfactorily decide'.

Mining Institute of Cornwall medal awarded to Holman Brothers and James McCulloch in 1881 for the Cornish rock drill.

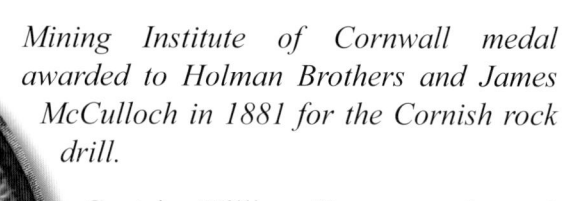

Captain William Teague, undaunted by his experience with the Beaumont drill, immediately ordered Cornish rock drills for Carn Brea where four of them speedily drove 7 fathoms in a month. This inspired his son William Teague junior to work on his own drill, the Cornish Boy, which was used with his patent ventilator in several mines.

Part of the original Memorandum between Holman Brothers and James McCulloch regarding the Cornish rock drill.

The success of the Cornish rock drill led to another agreement between McCulloch and the Holmans, which allowed them the costs for developing the prototypes of the 3 and 3¼-inch drills and another £84 towards maintaining the drills which Captain Teague put into Carn

Reverse of the 1881 medal.

The 1882 Memorandum between Holman Brothers and James McCulloch.

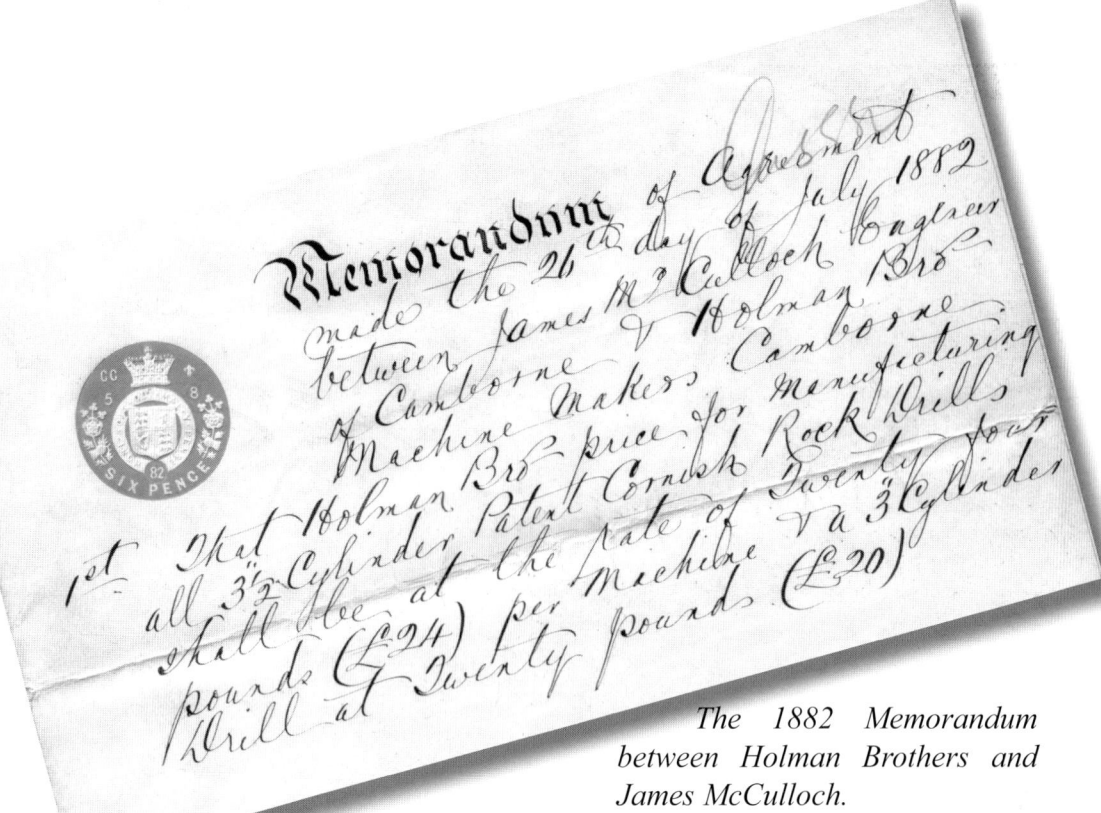

Brea. Yet there was the first intimation that the partnership was in trouble; clause 6 stated 'That as soon as Holman Bros, receive payment of any account arising out of the sale of a Drills.... they shall pay James McCulloch half of his half share until such time that he has paid back the money advanced on his behalf for Patenting Advertisements, Printing etc. when he shall receive the whole of his share'. A year later another rock drill contest was hosted by the Royal Cornwall Polytechnic Society when they had to bore into a large granite boulder supplied by John Freeman & Sons of Penryn and a very hard tin stone, full of 'vugs' (cavities) from Dolcoath Mine. The Cornish drill went up against not only its old rivals, the Eclipse, the Excelsior, the Barrow and the Beaumont but also a new machine, made by R. Stephens & Son of the South Crofty Iron Works, alias the 'Rock Drill Hospital', or as it was better known in later years, the Climax Rock Drill and Engineering Company. The two-day trial was dogged by poorly tempered drills and machine breakdowns; Captain Josiah Thomas considered it was more of a test of drill steels than the drills.

Nobody was satisfied and soon old rivalries reached the letter columns of the *Mining Journal*. Enraged by the claims of an 'Occasional Correspondent' who thought the Eclipse superior to all other drills, James McCulloch defended the performance and quality of the Cornish rock drill. The two feed screws driven by a central cog wheel ensured the machine stayed in balance on the cradle without wear or

Holman-McCulloch agreement of 1882:

For all Rock Drills supplied to J McCulloch Manchester I James McCulloch do hereby agree to forfeit the sum of Thirty pounds per machine the same to be deducted out of my share of the profits arising out of the sale of the "Cornish" Rock Drill provided that he does not pay as per agreement

Signed this 31st day of July 1882

James McCulloch

Witness Francis John Harvey

Mining Institute medal awarded to Holman Brothers in 1882.

jamming and the machine used half the amount of air consumed by the Eclipse. He roundly attacked his old partners, claiming that Hathorn & Co. were so worried by the new competition that they consequently 'fall back on the pen to what they fail to accomplish by fair contest'.

ROCK DRILL CONTEST AT DOLCOATH, December 1881

	Diameter of cylinder	Time boring	Depth	Cubic inches of Ground cut per minute	Mean pressure per square inch
"THE CORNISH"	3½ in.	2 min. 10 sec.	13½ in.	16.4	61 lbs.
"ECLIPSE"	3½ in.	2 min. 35 sec.	11¼ in.	13.6	60 lbs.
"BARROW"	4 in.	2 min. 15 sec.	8¾ in.	9.3	60 lbs.

ROYAL CORNWALL POLYTECHNIC SOCIETY
ROCK DRILL CONTEST, September, 1882

	Diameter of Cylinder	Stroke	Diameter of Bit at commencement	Depth bored in inches	Capacity of Hole in Cubic inches	Air consumed at 60 lbs. pressure cubic feet	Cubic feet of air at 60 lbs. pressure per cubic inch of Stone	Kind of Stone	Position of Hole	Depth of hole bored per minute	Cubic inches of stone cut per minute
HE CORNISH	3½	5	1	28½	64	230.4	3.6	Granite	Vertical	3.87	8.7
LIMAX	3½	5	1⅞	26	65	295.8	4.55	Granite	Vertical	3.06	7.64
XCELSIOR	3⅜	4	1⅞	27¼	65½	267.2	4.07	Granite	Vertical	2.72	6.55
CLIPSE	2½	5	1	22½	29½	152.0	5.15	Granite	Vertical	3.85	5.06
EAUMONT	3	5	1	5¼	12¾	90.6	7.1	Granite	Vertical	1.75	4.25
HE CORNISH	3½	5	1⅞	13	13	121.9	3.9	Very Hard Green Stone	Diagonal	2.43	5.85
LIMAX	3½	5	1⅞	10	10	234.1	9.06	Very Hard Green Stone	Diagonal	1.0	2.53
XCELSIOR	3⅜	4	1⅞	4¾	4	124.1	9.48	Very Hard Green Stone	Diagonal	.79	2.18
CLIPSE	2½	5	1¼	16	16	244.9	13.7	Very Hard Green Stone	Diagonal	1.6	1.77

Results of the 1881 and 1882 rock drill contests.

Hathorn angrily rebuffed McCulloch, claiming that he imagined that 'by maligning the Eclipse drill, he can enhance the value of the Cornish drill but we think differently' and that 'it was us, Hathorn & Co. who first introduced Mr McCulloch into Cornwall and found him the machinery to take his first contract'. They ended by likening him to the exhausted snake which, on being taken home by a man, then bit its benefactor.

'Occasional Correspondent' returned to the fray claiming that the Cornish rock drill was over-engineered, over-manned, and that on the day before the contest it had broken down at East Pool Mine and been replaced by an Eclipse. The Holman boys joined the argument over the contest and defended their drills in East Pool, saying just one machine had been out of action for only a couple of hours. The Cornish drill which Captain Teague had used underground in Tincroft for the last nine months had never been back to the fitting shop. This was the same machine which beat the Eclipse in the trial in December 1881, and anyway the Eclipse drills tested recently in Dolcoath used so much air that it was known as the 'consumptive valve drill'. Michael Loam, the resident engineer at Dolcoath who was building the Barrow drill under licence, joined the fray when he sent out with a sales leaflet, a distorted copy of a letter by Frederick Doering about his drills' performance at Rio Tinto. The McKean Company, which was driving the St Gothard Tunnel, then offered to take on all comers and the Excelsior drill makers and the Holman boys accepted. James McCulloch continued to stoke up the general resentment but in the end there was no contest.

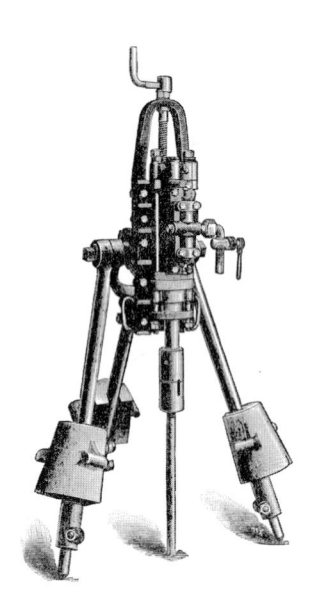

The Scrann rock drill of 1907 - the design unchanged despite 30 years of development.

There was also increasing friction between him and the Holman boys, unsurprising as they belonged to the Camborne mining and engineering fraternity, notorious since the days when Boulton and Watt desperately tried to get royalties for their new engines. A breach was inevitable when McCulloch continued to object to the high cost of advertising. He insisted that they had agreed adverts would be confined only to the pages of the *Mining Journal* and besides, when these did appear, Holmans' general mining machinery was much more prominent than the Cornish rock drill. He also took umbrage at the way the Holmans took rock drills to contests and exhibitions without consultation and then sent him the bill for half the costs.

James McCulloch returned to Huelva and Rio Tinto but was outraged by the accounts that the Holmans sent him, suspecting that though the brothers were not

Part of an early Holman Brothers parts list, 1881.

actually dishonest, they were offering discounts and doing deals with their friends in the mining world. Finally certain 'that all these irregularities go to show the loose system and vexilating methods Holman Brothers have' he decided to sue them and they promptly counter claimed for £100 17s 5d. The acrimonious dispute was not settled until 13th August 1887 when they agreed to cancel all debts and each could develop the patent for the Cornish rock drill, free from interference of the other. The Holmans were still not taking chances with McCulloch and had soon secured their own rock drill patent to cover the Spanish mines.

Even with the disgruntled McCulloch in Spain, the Holmans managed to keep up a brisk trade even from the early days of rock drills, as the following items from an old Day Book show:

The Camborne works in the 1890s. Much of this layout could still be recognised when the site was demolished in 1994.

Oct. 25th 1886 – 4 3½" drills with accessories and equipment

Apl. 18th 1888 (Peñha) – Air Compressor, Boiler, Air Receiver, winch and trolleys, 2 3½" rock drills, mountings and equipment

Apl. 23rd (Peñha) – Air Compressor with Boiler, 1 3½" drill with mountings and accessories

This era of frantic and often bitter rivalry between the rock drill companies gradually ebbed away as technical development evened out the markets. There was a final flurry in the pages of the *Mining Journal* following an exhibition of mining held at the Crystal Palace in May 1890. James McCulloch had the largest display of all and his less than honest claims that the Cornish rock drill was merely a crude prototype, whose imperfections enabled him to perfect his magnificent new Rio Tinto rock drill, must have outraged the Holman boys. They were also assailed by Climax which disputed Holman claims as to the merits of their valves, durability of machines and especially that they had invented the steel liners which took the wear on the cradle as the drill was wound back and forth. Climax claimed to have done this years before but had strangely not patented the improvement. They made the inevitable challenge to a contest at £400 a side to split between the Miner's Hospital at Barncoose and a London charity. The Holman boys, although successful in such contests, remained dubious about them and told Herbert Thomas of the *Cornishman* newspaper that they believed that in a trial 'lasting only a few minutes, a drill may be dismissed as worthless on account of a borer happening to be too hard or too soft although there may be no inherent defect in the drill', and anyway 'We prefer to stay at home and execute orders'.

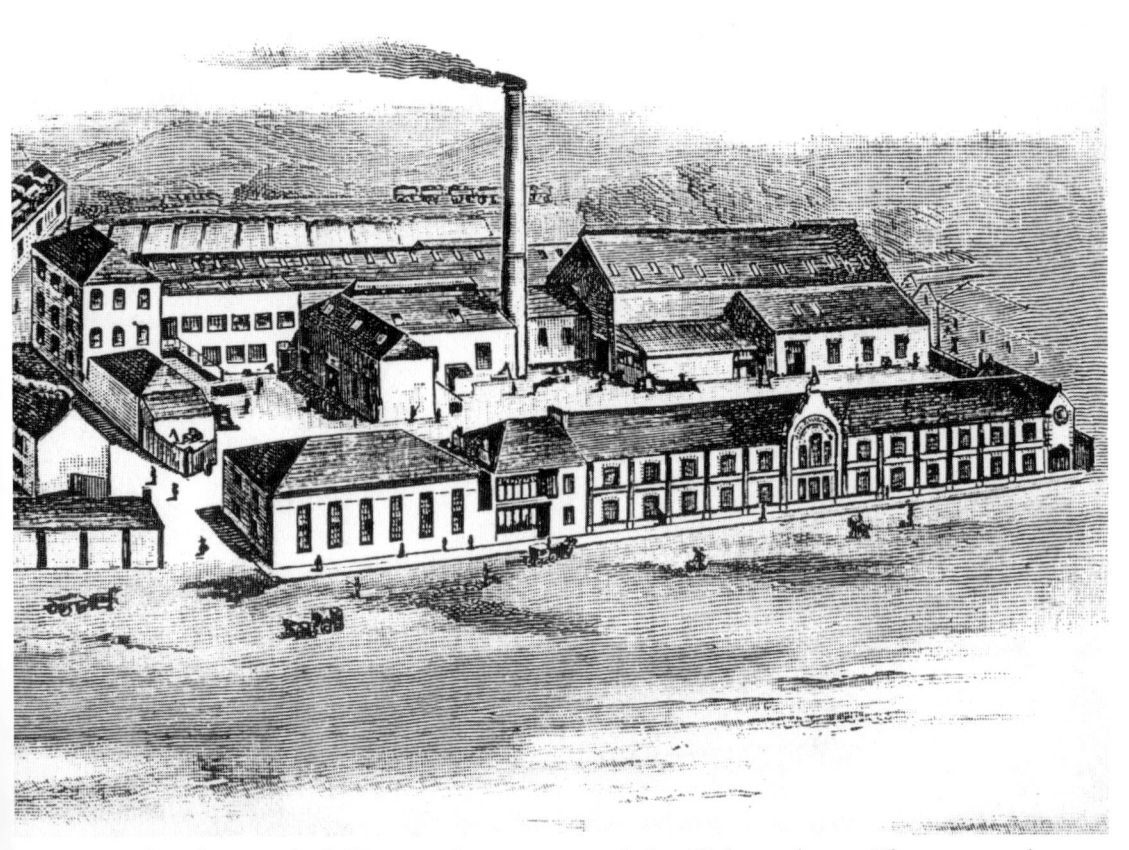

They withdrew from the argument when their father John died suddenly at the age of seventy-two while attending a missionary meeting at Camborne Wesley

Late 19th century view of Dolcoath Mine, one of Holman Brothers main customers.

on 6 November 1890. He was really the last link to the early days of steam engineering and the great copper mining era, who heard tales of the great Trevithick from his own father. On the day of his funeral the Dolcoath adventurers closed their meeting so they could join other mining men and Holmans men in a long procession through Trelowarren Street where the business closed as a mark of respect.

Transporting a boiler from factory to mine.

CHAPTER THREE

'SUCCESS TO THE COMPRESSOR'
(Major White of Levant Mine)

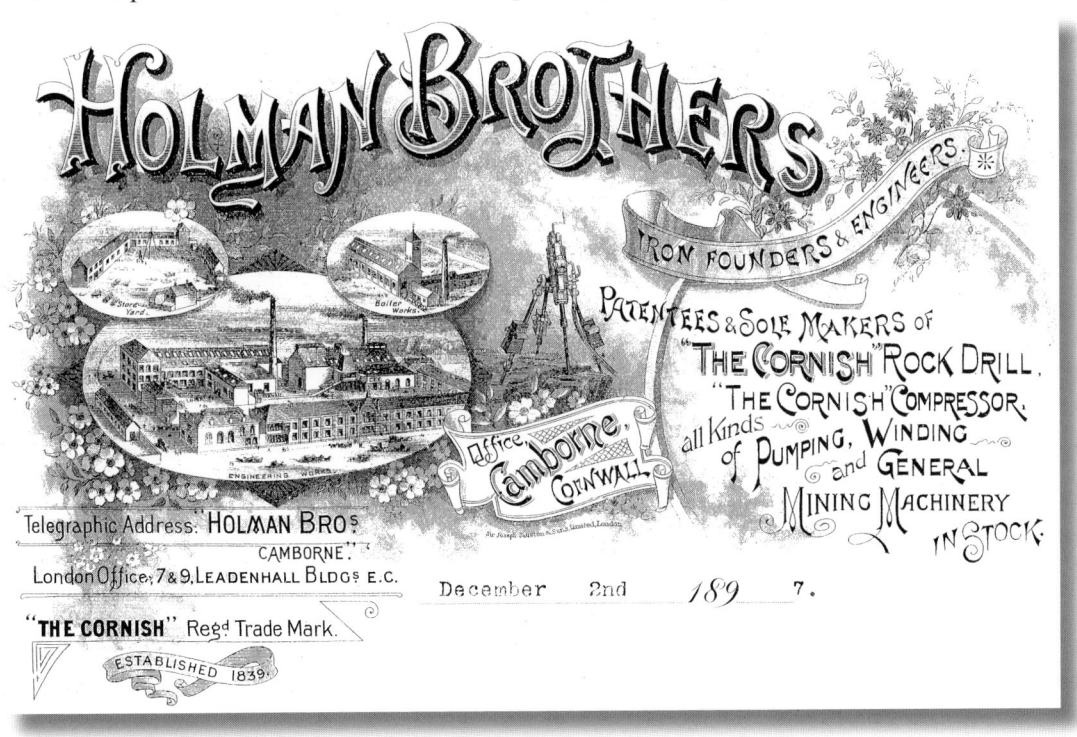

John Holman had lived long enough to see his sons create a thriving engineering works and foundry. In January 1887 they erected a big new air compressor for Dolcoath and later that same year the last ever Cornish beam whim at East Pool. Known as North Whim, but more often as 'Michell's Whim' after its designer, the Cornish engineer F. W. Michell, the 30-inch cylinder engine worked at 25 strokes per minute and could haul skips at 1,000 feet per minute. At the same time they built a 16-inch engine with condenser and pitwork 'with a thoroughly good whistle, to be heard 2 miles off' for Mr Strauss. There were also plenty of foreign orders:-

Dec. 30, 1884 - One 3″ and one 3½″ drills and stretcher bar. 24 bits, 1 spare valve tappet and pin for each, and clamp. For Spain.

June 18, 1885 - Two 3″ rock drills, Governor and spanners. Stretcher Bar and clamp, with hose, borers and spare parts ("duplicates") For New Zealand.

June 20, 1885 - Steam Winch. One 3″ Drill. One 3½″ drill. One Stretcher bar. For Australia. 25 October 1886 - Four 3½″ drill, 2 handles, 2 sets spanners, 4 clamps, 2 Governor cock, Pistons as per tracing For Spain

An old foundry day book tells of the work that went to fulfil these orders;

9 July 1888 Fitting shop. Ben Cock, Bond, G. Hashing and T. Penhall. To make 2-3 ½″ drills. Thirteen days.

13 August 1888. Ben Cock and Bond to do all the turning including twist bars of 2 - 3 ½″ drills commenced at noon.

22 Feb. 1889. H. Harvey and Jackson to make, try and paint one 4" winch. 14 days. Cylinder bored. Covers turned.

Ben Cock, J. Hashing, T. Penhall and F. Miller to make four 2 ½" drills. 13 August 1888 Ben Cock and Bond to do all the turning including twist bars of 2 3 ½" drills. Commenced at noon.

Eighteen year old Ben Cock was one of one of five talented brothers of an old mining family whose father was known as the 'Solomon of Tuckingmill'. Ben's brother William 'artist, draughtsman, author and one time preacher' had, like Ben, joined Holmans in 1878 and was responsible for the magnificent artwork for the company publicity for over fifty years. Their nephew was Garstin Cox, who matured into a superb artist who would produce paintings for the famous Holman calendars. Ben Cock was a dedicated young man, eager to improve himself by study and hard work. His subsequent career was inseparable from the development of the rock drill and he rose to become manager of the No. 1 Works.

Holmans had already set up their own small rock drilling company to carry out

Group from the carpenters' and pattern shop, early 1897.

Back Row: Josiah Pendray, W. John Ward, Alfred Cock, Frank Godolphin, David Vellanoweth

Second Row: John Jennings, M. Mill Trevarthen, Henry Pappin, J. C. Mennear, Robert Wheeler, H. A. Miller

Third Row: G. B. Tangye, James Retallack, Nicholas Curry, Alfred Harvey, Tony Harvey, John Menhennet

Front Row: Aldrovand Maynard, Arthur Angove, Thomas Webb

driving and sinking contracts for the Camborne mines. They drove at Wheal Seton where the venerable Captain Charles Thomas, father of Captain Josiah of Dolcoath, on his first visit to the mine for thirty years, was deeply impressed by the Holman machines.

Holmans also executed what was regarded as an amazing feat of machine drilling in a Cornish mine. During the summer of 1889 Captain Bishop had brought back rock drills to Wheal Grenville

In 1896 the Pengegon Steel Works was employed to make drill steels for Holman Brothers; this medal was struck to commemorate the event.

45

after the miners objections had led to the pulling out of three Champion rock drills some years before. He also chose to improve haulage and winding at the old East Grenville Engine Shaft by simultaneously sinking and rising and Holmans took the contract to put up the 51 fathom raise at £31 10s a fathom. 'Raising' was an unpleasant job, with the drills boring almost vertically upwards above the heads of the miners, who faced a continual cascade of dust, dirt and muddy water, perched as they were on a box staging which was accessible only by ladders that disappeared into the gloom beneath their boots. There was also the pounding roar of these early drills which, weighing over 300lb, had to be dismounted once a round was drilled and manhandled out of the way of the blasting.

Far above the Holman crew's heads, the hand labour men were sinking the bottom of the shaft until in October 1890 they broke through. The engine shaft had been risen 30 fathoms by the Holman crew, and such was the accuracy that it was perfect on the south-west corners and only a matter of inches on the north-east. The Holman boys' delight was somewhat marred when an expatriate Cornish mine captain, W. J. Paul of Charter Towers, the big Queensland gold mine, having noted the controversy between Holmans and Climax, wrote a scathing

Holman Brothers moulding shop, about 1900.

letter regarding the Grenville rise. Holmans might rise at 21 foot a month but at Charter Towers, with four Ingersoll drills they put a shaft down 50 feet in a fortnight. The Holmans ignored the potential controversy but James McCulloch, back at Huelva, quixotically rose to their defence, or rather to that of the drill he helped design, in the pages of the *Mining Journal*.

It was about this time that the old boiler works that John Holman had originally established at Pool was moved to Roskear, on part of North Roskear Mine. While it would have different names in the future, such as The Cornwall Boiler Company and The Cornwall Drop Stamping Company, it would always be known as 'Boiler Works'.

Group of Holman workers in 1903.

Meanwhile the Holman works continued to expand and its business largely escaped the effects of the catastrophic tin slump which almost wiped Cornwall off the world's mining map in 1896. The *Mining World* noted during a visit in 1894 that 'the major part of what is produced is despatched to foreign mining fields... seven eighths of the work turned out here is on order for all quarters of the globe. Some of the workmen were engaged in moulding a girder-type bed for an 18-inch by 36-inch two stage air compressor.... There is on view Holmans' improved pumping and winding engine for steam or compressed air. The fitting shop is one of the largest in the West of England - 200 feet by 40 feet wide. Here were found some thirty men employed on overtime making Cornish rock drills for delivery in various parts of the world'.

It was around this time that a woman in Troon developed a method of feeding

Parts list for the 2½-inch stoping drill.

the various Troon men who worked at Holmans. Every morning, around eleven o'clock, she would go round the village with a donkey and cart. Every few yards she would blow a whistle, a signal to the women, who would bring out pasties for croust or crib. Down at the works it would trundle around each section and the men and boys would search for the pasty with their name on it. Presumably a charge was made for the service.

Another big compressor had been built when the old committee at Carn Brea Mine, soon to amalgamate with neighbouring Tincroft, realised that their old compressor was burning 10 tons of coal a week merely to work two Cornish rock drills sinking Old Sump Shaft. This was 3 tons more than the Tincroft air compressor needed to work double the number of drills and two Holman air winches. The Holman boys responded with a technological monster that marked the real beginning of their long tradition of building air compressors. Installed in the spring of 1894, it was a cross-compound compressor with a high pressure cylinder 18 inches by 72 inches and a low pressure cylinder 34 inches by 72 inches driving an air cylinder 26 inches by 72 inches, while the flywheel alone was 15 feet in diameter and weighed 14 tons. Capable of driving fifteen Holman drills and three 6-inch combined pumping and winding engines, it was 75% more efficient and burned 1½ tons of coal less per day than the old machine.

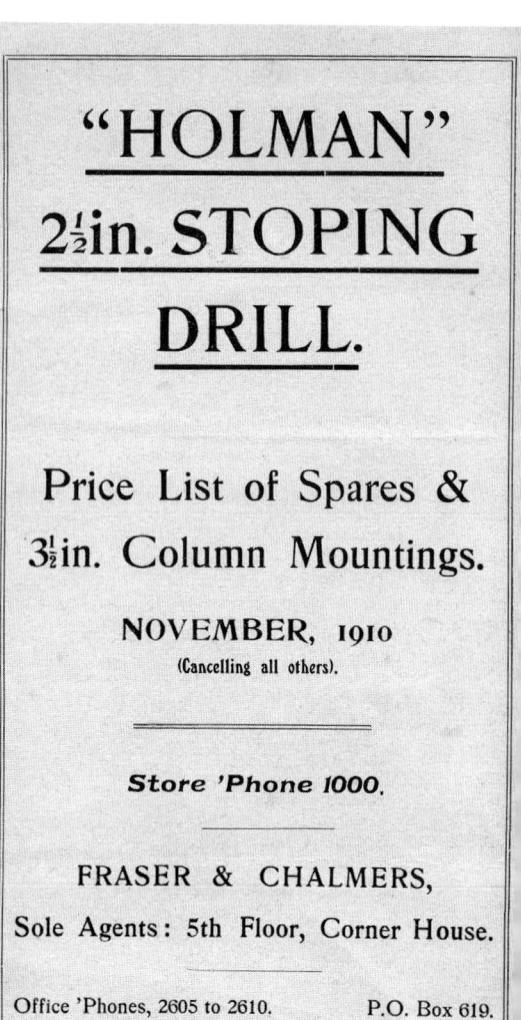

There was a spectacular tribute to Holman workmanship when, on a fine night in May 1904, P. C. Meneer and engineer Rodda found the compressor house engulfed in flames with the roof about to collapse. To everybody's amazement, including Nicholas Trestrail and Captain Penhall, the blaze had been so swift that, apart from broken steam gauges and slightly scorched paintwork, the compressor was unharmed. Luckily the fire brigade had not been called as cold water turned on the hot iron work would have been disastrous for the engine, which was not insured. However, months later the air cylinder was found to be cracked which stopped the compressor working for some days.

The great Dolcoath Mine was also modernising, though the ailing Captain Josiah Thomas had largely handed over management to his son R. Arthur. Arthur already had wide experience of mining, having been manager at City & Suburban Gold Mine in the Transvaal for some time. His enthusiasm for mechanisation exceeded that of his father for the 'boring machine' and the mine was to undergo radical transformation above and below 'grass'. Unsurprisingly, Holmans would do much of the work as they were both excellent engineers and large shareholders. John Henry Holman sat on the board of the limited liability company formed on 2nd October 1895.

A few weeks later Michael Williams, the chairman of Dolcoath, cut the first sod of the new Williams' Shaft, which, once sunk 3,000 feet beneath Carn Entral would supposedly modernise the entire mine. Buffeted by a sudden snow storm, the assembled company retreated down the hill to the count house, and what an amazing assemblage they were, redolent of the wealth and power

The directors of Dolcoath Mine posing at the ceremony for the cutting of the first sod of Williams' Shaft.
Rear: Allen Stoneham, Josiah Thomas (Managing Director), Frank Harvey.
Front: George H. M. Batten, Michael Henry Williams (Chairman, with shovel), Oliver Wethered (Deputy Chairman), William Rabling.

that controlled the 'Black Country' of Cornwall. Among them were the Holman boys, Captain Josiah and three of his sons and Oliver Wethered, who would later create Geevor Tin Mine. Also there were Arthur Strauss, tin broker and Liberal M.P. for the Mining Division; Frank Harvey of Harvey's of Hayle and David W. Bain, mine adventurer and owner of a fleet of steam colliers. Another attendee was Major Francis Oats of St Just, a director of De Beers and friend of Cecil Rhodes. He was a member of the Cape Parliament and had returned home to invest his wealth in Cornish mines. George H. Batten, barrister and director of the Hyderabad Deccan gold mines which had just bought a large air compressor from the Holmans, was also there.

James Miners Holman himself proposed the toast 'The Neighbouring Mines', remarking that he hoped they would follow the example of Dolcoath which 'had broken away from the old lines and started on a new foundation'. He was sure that young Arthur Basset of Tehidy would be equally as generous to them in offsetting their extra costs, and he was heartily supported by Captains White of Carn Brea, Teague of Tincroft, Hambly of Agar and Bishop of Wheal Grenville.

Dolcoath's old man engine was increasingly regarded as a lethal and obsolescent contraption. Installed back in 1854 and capable of carrying 108 men on a twenty minute journey almost

The mighty winding engine built for the Basset Mines.

to the bottom of the mine, it had once nearly killed Captain Josiah himself. Holmans had repaired the main beam which broke in 1893 and shortly before being stopped in October 1897, the man engine exacted a final toll; one miner

Holman catalogue engraving of the huge traversing winding engine for Williams Shaft (see overleaf).

was killed in the morning and another seriously injured during the afternoon. Henceforth men would ride in man cages at Harriet Shaft where a new steel headgear and a Holman air compressor had been erected.

Dolcoath's familiar Cornish headgear also vanished and by the spring of 1896 a new steel colliery-type headgear towered above New Sump Shaft. Built by Holmans to the designs of Charles Morgans of Cardiff Mining consultants T. W. Morgans who were now consulting engineers at Dolcoath, it was erected in less than a month. They scrapped the old Cornish beam whim that hoisted from New Sump Shaft and replaced it with a small horizontal Welsh colliery winder made redundant by the closure of Botallack. After inspection by R. Arthur Thomas, it had been bought from the Holman cousins and brought to Dolcoath behind a traction engine.

Detail from a Holman day book from 1906. It features the well-known names of Nicholas Curry, Ben Cock and Henry Nettle.

Another new steel headgear was erected on Old Sump Shaft while there were more developments over on Stray Park, where Holman contract sinkers were putting down the shaft and driving levels despite very hard ground. Holmans also refitted the ancient Stray Park 64-inch pumping engine which, having been re-bored once by Harvey's, was now re-bored again and fitted with new gear and boilers.

Perhaps unnerved by this great wave of modernisation Michael Williams, famed for his remark that he failed to see 'What dividends had to do with profits', finally resigned as chairman of Dolcoath after forty years. Frank Harvey, chairman of Harvey's of Hayle, was elected in his place and the vacant seat was filled by James Miners Holman who joined his brother on the board.

*The Holman-built lattice headframe over Williams' Shaft on Carn Entral.
The winder house is to the right.*

The new Williams' Shaft was beset by problems from the outset as the weather remained bad for months. Doubts increased that the shaft was far too big, especially when the sinkers struck granite that was either very hard or badly faulted and struggled to cut the shallow adit to help drain the sump. Work was suspended for some time and when it resumed it was obvious that a replacement was needed for the small sinking winder supplied by John Wild & Sons of Oldham in 1895. The old style Cornish beam whim, such as Holmans had erected at East Pool in 1887, would have been enormous and cumbersome, especially as any headgear and machinery would be crowded together on the steep side of Carn Entral.

There was the additional problem of keeping the vast length of wire rope coiled centrally on the winding drum while hoisting from such a depth but a 'daring and original' solution was offered by Charles Morgans, of the company that worked on the modernisation of Dolcoath. The engine and drum were mounted on a girder bed which itself rode on 16 paired bogie wheels travelling on steel rails, set at 90 degrees to the shaft. The whole 120 tons moved 16 feet laterally for every 1,000 feet of wire rope, which always remained coiled centrally over the drum.

In January 1900 an accident was recorded from the foundry. One of the smiths, James Lanyon, caught his hand under the steam hammer, which badly damaged the fingers. After being taken to the Miners Hospital near Redruth he had four of them amputated.

Holmans undertook the work, including the 60 foot lattice steel head gear and by February 1900 the sinking crews were ready to begin. However, although the engine 'was being erected and the pit head gear ought to be very soon' James Holman admitted they were late as strikes up country had delayed material for the foundry. Captain Josiah Thomas was absent, this being the first Dolcoath meeting he had missed in thirty five years, but he wrote to recommend that they have a new air compressor capable of driving 20 drills. This was more than he required but guaranteed the power needed to develop the mine westwards towards Stray Park.

By August 1901 the new winder had been installed and tested and the wire ropes were being wound onto the drums and, according to Frank Harvey the chairman, in 'two or three days the contractors would be at work in the shaft sinking rapidly'. A model of the winder had been made by Holmans for their stand at the Paris Exposition in September 1900, to show the principal features of the engine (This model was discovered in a number of pieces at Poldark Mine, Wendron, where it is being rebuilt). There they won two medals for their display of rock drills which were already in use on the South African gold mines. Also on their imposing stand

Holman Brothers' rewards from the 1900 Paris Exposition: Bronze Medal (above) and Gold Medal.

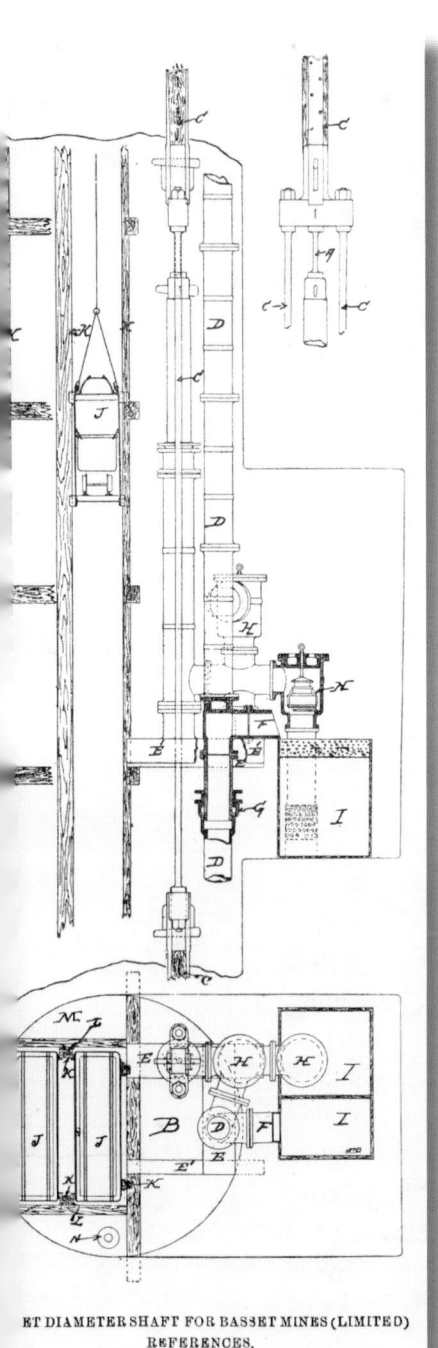

Nicholas Trestrail's diagram of the pumping and hoisting arrangement in Marriott's Shaft at the Basset Mines.

was a 'Duplex' pump, a portable headgear used widely in Australia and West Africa and a patent filter able to supply water for drinking or a boiler. George Batten, the Dolcoath director who visited the exposition, noticed that the model 'caused great interest and curiosity amongst people who included many high officials and who expressed a wish to see it at Dolcoath when it was in full work'. An extra visitor to the Exposition was seven year old Arthur Trevena Holman, always known in later life as 'Mr Treve', who travelled across under the care of Mr Smith, who would become the celebrated manager of Holmans' London office.

Williams' Shaft never really fulfilled expectations and when temporarily suspended again through a shortage of funds in 1905, was even then being referred to as an 'expensive white elephant' by the *Mining Journal*. R. Arthur Thomas was always reticent about developments at the shaft and an indecisive discussion over the pumping, and this led J. H. Collins to claim that questioning him had as much effect as 'a corn plaster on a graven image'. When sinking was resumed, Dolcoath, much to Holmans chagrin, hired Piggott's, the Welsh pit sinkers who used the German Flottman drills which Holmans had regularly beaten in contests in South Africa. The shaft finally bottomed out at 3,000 feet in May 1910 at a cost of £90,362, which included the remarkable winder.

The race for modernisation brought Holmans further customers when the run of old mines along the Great Flat Lode, between Troon and Carnkie, amalgamated as the Basset Mines in January 1896. The managing director was Major Frank Oats of St. Just who, like R. Arthur Thomas of Dolcoath, had been converted to mechanised mining by his experience in South Africa. He needed a new winder to hoist from Marriot's Shaft, where it would work alongside an enormous 40"/80" compound engine which, he hoped, would cope with the huge amounts of water that daily needed to be pumped from the workings. This compound engine was designed by Henry Davey, an engineer of Cornish descent, who was working for Hathorn & Davey's 'Sun' Foundry in Leeds at the time. Nicholas Trestrail, the Cornish consulting engineer, issued tenders for a winder to hoist double skips raising 1,000 tons a day from a shaft that might reach 5,000 feet. Holmans beat off fierce competition and, for £3,600, produced a horizontal compound winder which 'incorporated all the modern appliances for speed, power and efficiency'. It had cylinders of 23 inches and 38 inches, a stroke of 5 feet 6 inches and was equipped with fast-acting 'Corliss' valves. The winder also had massive conical winding drums, themselves an innovation in Cornwall. Holmans also erected the massive steel headgear, which cost three times as much as the old Cornish type headgear that was already fast disappearing.

James Miners Holman proudly informed the *Mining World* that he was 'making the biggest compound winding engine that has ever been put up in the county' when they

Holmans' patent compressor with the fast-acting Corliss valves.

Post card showing the pumping and winding arrangements at Marriott's Shaft at the Basset Mines. The winding engine is on the right; the compound inverted pumping engine is behind the head frame.

interviewed him at Holmans' new London office where Arthur Herbert Smith, formerly an estimator at Camborne, was the representative. Since they had spoken to him, the works had doubled in size to cope with orders for rock drills and mining machinery. The boiler works had been expanded and the moulding shop extended. He had also taken over the old foundry near the railway station and was converting it into an apprentice training school where, under the instruction of skilled men, the lads would be taught to make small engines and light work before graduating to the main works.

The London Office, or 'L. O.", at Broad Street House had been opened in 1898. Smith had only joined the company from the Mid Cornwall China Clay Company the previous year in 1897 and his promotions were nothing short of meteoric.
The new winding system at Basset Mines was not infallible, as on one Christmas Eve everybody 'was rushing to leave work early and the driver steamed the engine a little too far. The cage with two full wagons went up the headgear and broke off at the chains...'. The cage dropped back down the shaft, travelling so fast that in the dark that no one saw it, and was eventually found 260 fathoms below, having sliced through eight feet of timber, crushed into the sump where it stayed until sinking was resumed years later.

Holmans made a similar smaller hoist for the restarted Gooninis Mine near St. Agnes but the next real masterpiece came after Major Oats returned with another order. As well as having considerable interests in the Basset Mines, he was also one of the largest shareholders at Levant. At the turn of the century this mine was still notorious for its obdurate, often eccentric management and obsolescent equipment. He had campaigned to modernise the mine and a big coupled air winch was installed on the 260 to hoist from the deepening levels. The old Harvey compressor hidden in a shed behind the engine house was inadequate and in June 1899 Captain White insisted that they 'must have a new compressor if they were to open up the mine', while the miners 'believed they would be able to do one third more work if they had air at higher pressure'.

The Holmans were already assembling a massive cross-compound stage compressor for Dolcoath. It had Meyer valve gear and an 18-ton flywheel, the 'largest ever made in the moulding shop'. This would enable R. Arthur Thomas to centralise the air supply, which relied upon five compressors scattered between Wheal Harriet and Valley Shaft. It would also save the mine 100 tons of coal a month. Without much prompting from the Holman cousins at St. Just, the Levant committee went to Holmans, who were determined that the new compressor would be the 'best piece of machinery that had ever been turned out of their

The 2,000 cu. ft. compressor installed at Dolcoath.

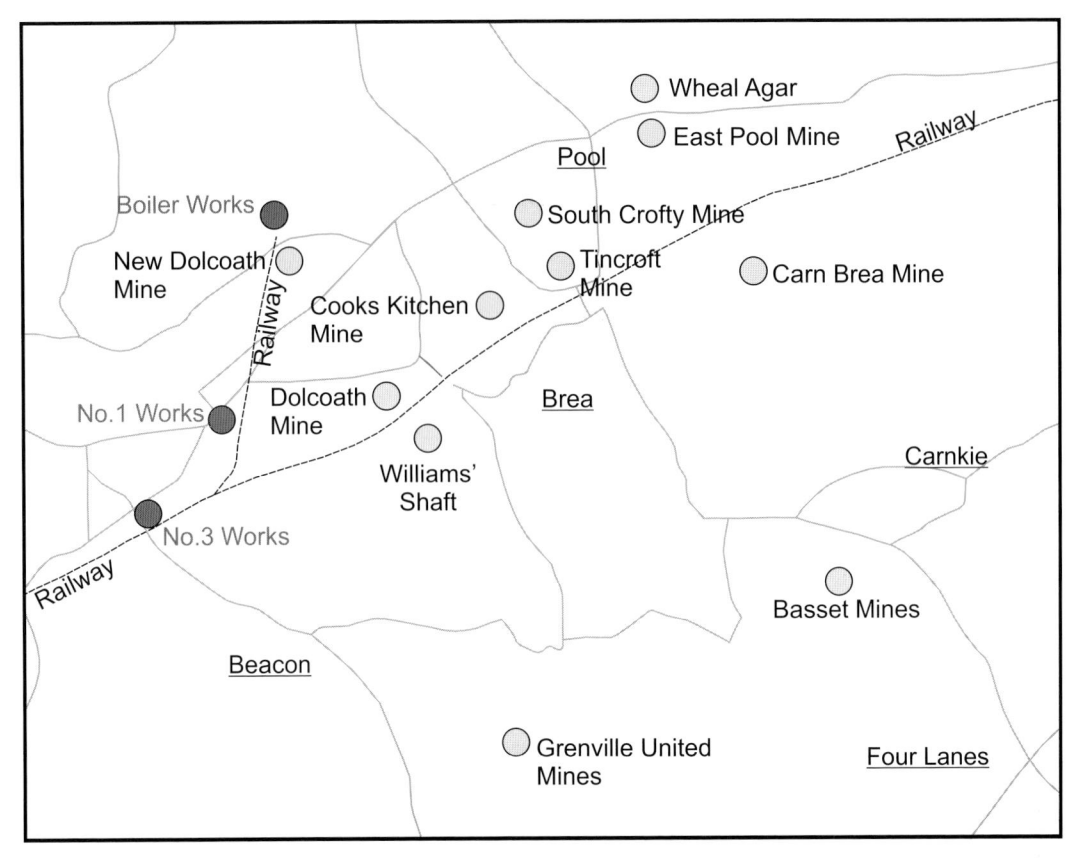

Location of the three Holman works and the local mines.

works'. It was double the size of those supplied to Carn Brea and the Hyderabad Mines only a few years before and, unlike the new one for Dolcoath, was driven by a triple expansion jet condensing steam engine equipped with Meyer steam expansion gear. The tandem air cylinders could draught over 3,000 cubic feet of 'free air' per minute and supply through the receiver 20 rock drills at 80 psi. The compressor was over 60 feet long and 16 feet wide, with steam being supplied by two Lancashire boilers.

While the foundry put their best energies into making the enormous compressor, John found the opportunity to begin his palatial new villa at *Camborne Veor*, which lay midway between No. 3 Works and his brother's house at *Tregenna* in Pendarves Road. He was already a widely respected businessman and J. P. and soon his brother James was also sworn in as a J. P. at Bodmin. By June, James was travelling across Canada and the USA to seek business in the mining fields of British Columbia, accompanied by Captain William Bennetts of the

Roskear Fuse Works.

All work ceased on 26 October 1901 for the funeral of Captain Josiah Thomas at the Centenary Chapel 'from whose pulpit of which his voice was so often heard'. Ten thousand people lined the streets as his coffin was carried from the mine, borne by the six captains at Dolcoath. The cortege was led by Frank Harvey, the chairman, Gilbert Pearce and James Miners Holman. They were followed by the clerks and time keepers and Thomas Morgan of the Welsh engineering company, then 800 tin miners, marching silently and respectfully. It was an extraordinary event, never accorded to any other man in Cornwall, and marked the passing of the last great figure from Cornwall's golden era of hard rock mining.

The christening ceremony for the new Levant compressor, which took place on 22 December 1901, was reminiscent of those once reserved for the starting of some big Cornish pumping engine. To escape a typical wet and stormy day on the Pendeen cliffs, everybody crowded into the magnificent house where they were shown over the compressor by John and James Holman, Nicholas Trestrail and Mr Plummer, Holmans' Chief designer. Major Richard White, the doyen of Levant, gallantly offered the honour of starting her up to Major Frank Oats and the compressor 'went off as smoothly as a clock and was the subject of general admiration and satisfaction'. Toasts were drunk in the special punch brewed by

The Holman compressor at work at Levant.

Staff at the No.3 Works in 1905; note the large proportion of young people.

BACK ROW (Left to right): "Cocky" Vivian, Jimmy Oliver, Jack Udy, Tom Martin, Joe Rodda, Mark Trebilcock, Alfie Bray, Frank Blarney, Jack Tromans, Bert Turner, Bill Daddow, Sid Roberts, Jack Berryman.

SECOND ROW : Arthur Thomas, Jack Quick, George Richards, George Bawden, Jack Reed, Billy Gilbert, Ben Cock (Works Manager), F. C. Eddy (Clerk), Edward Arnold, Bill Combellick, Charlie Collins, Bert Tresidder, Howard Willoughby, ---------?, Dick Bennetts.

THIRD ROW : Bill Eade,----------?, Sam Osborne, Jack Holman, Wilson Uren, Willie Daniel, Willie Curnow, S. Dunn, CharlieTrevena, Bill Clemo, Henry Tonkin, Artie Rosevear, Edwin Bray (Foreman), Jack Vivian, Herbie Brown.

FOURTH ROW : Billy Brown, Johnny Crowle, Freddie Rule, Tom Matthews, Meto Trebilcock, Ernie Capel, Arthur Weekes (Office), Tommy Trevena, Jack Sowden, Sammy Williams, Charlie Vivian, George Chinn, Ernie Trewin.

Major White, who proposed 'Success to the Compressor'. Frank Oats was slightly overwhelmed by its sheer size and power, which was far more than was required at present and it was now up to the Levant agents to ensure they used its power to develop the mine.

John Holman claimed the compressor was faultless in every way, despite intense inspection by a bevy of engineers inside and out. James Holman was rather more lyrical, commenting 'a good compressor was like a good picture it needed a good frame and they had every reason to be pleased at the manner which the Levant company provided the frame'. He referred to the magnificent compressor house with its tiled floor and 100 foot ornate chimney with its Gothic concrete cupola. It comprised 2,000 tons of masonry which had cost a pound a ton, enraging

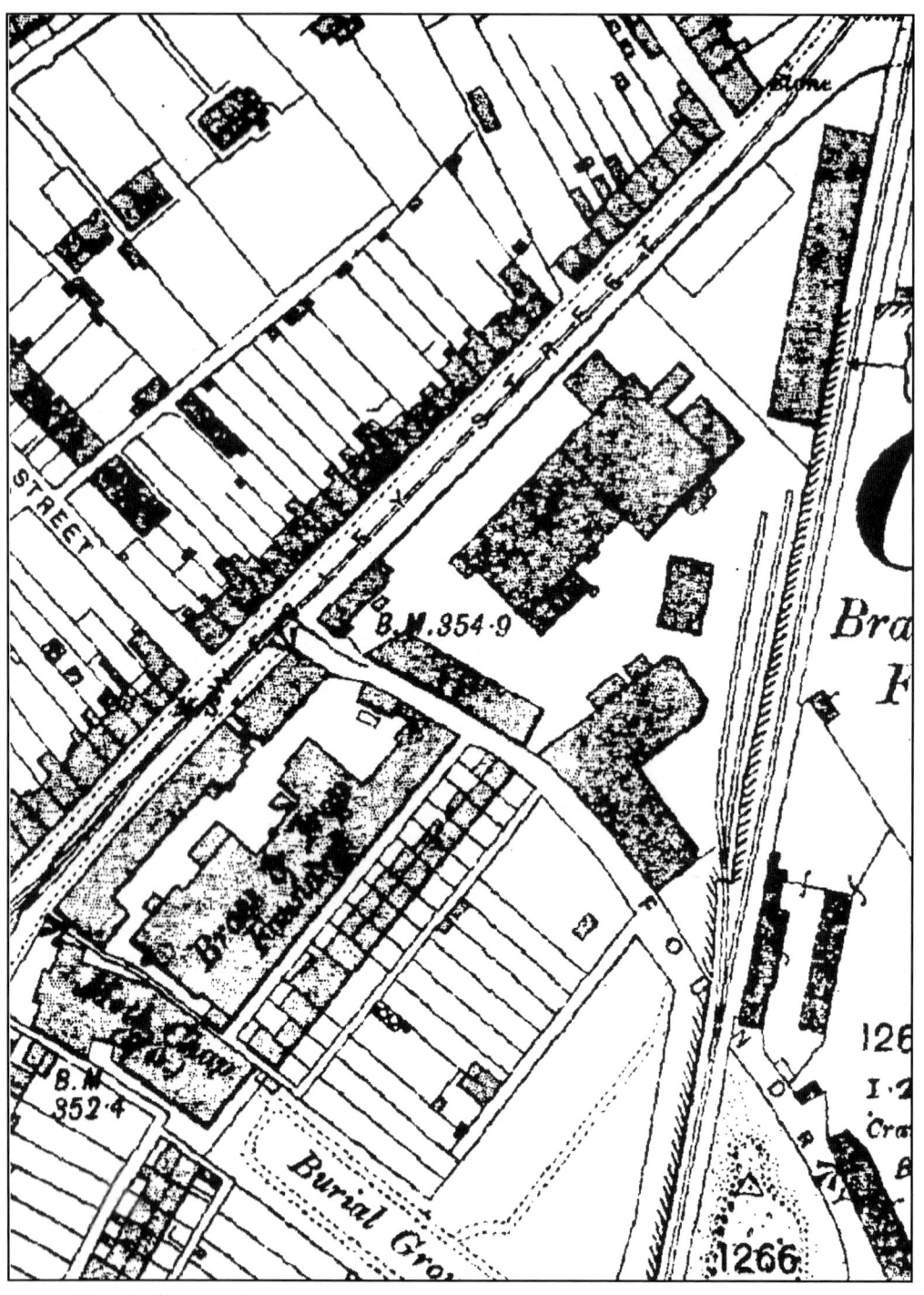

1908 Ordnance Survey map showing the great extent of the No.1 Works.

Head Office staff, 1910; the straw boaters may be evidence of an outdoor event.
Standing: A. C. R. Goad, Ledgers and Wages; J. H. Reynolds, Estimating and Shipping; A. M. Bray, Secretary's Office; W. J. Robins, Buying and General Shipping
Sitting: Matthew Curry, Accountant and Book-keeper; Wm. Edward White, Company Secretary; Nicholas Curry, Wesley Street Works Manager.

many shareholders who resented this expensive 'elaborativeness'. Indeed, when Nicholas Trestrail submitted his bill for erecting the compressor, the Levant committee reduced it by £17 and dismissed him.

George Plummer, Holmans' chief designer who had worked with Nicholas Trestrail, considered the compressor superior to any he had seen in Germany or America and he had worked 'with some of the best engines in the country'. With proper care and maintenance she would still be at work in twenty years, without fear of ending on the scrap heap. A new feature was noticed by Captain Williams, home from South Africa. On each side of the bed were cast the words 'Manufactured by Holman Brothers of Camborne'. The day ended with a response by Mr Holman junior which brought cheers, most from Treve Holman, who over thirty years later would be instrumental in saving the small steam winder at Levant for posterity. Sadly, George Eustice, the consultant engineer, had died on 3rd December 1901, only a few weeks before the ceremony.

The Holman boys were always proud of their roots in the Industrial Revolution and just a week after the Levant ceremony, John led the parade of traction engines, miners and engineers in celebration of the centenary of Richard Trevithick's road locomotive going up Camborne Hill on Christmas Eve 1801. He and J. J. Berringer, the Principal of the Camborne School of Mines where he was vice governor, spent months organising the parade. It began, on an appropriately wet and rough afternoon, from the presumed site of Tyack's smithy in Tehidy Road and ended in the main square, opposite the ornate fountain which John senior had erected in 1889. Amongst the scores of marchers was none other than James McCulloch, still the engineering nomad and about to quit Tuckingmill Foundry where he had been developing the Little Hercules rock drill for several years.

The Holman boys now employed over 100 men and were making 1,000 rock drills per year besides general mining equipment. Their contract rock drillers were kept busy, though occasionally this caused trouble in mines where they were also shareholders. An irate shareholder at East Pool, faced by heavy losses, asked why they had to employ Holmans when they could buy their own machines. John Holman brusquely told him that '..as long as his firm were makers of rock drills, they would be contractors... We have been asked to contract in South African mines, again and again.... Whether you contract again in East Pool is immaterial to us, because we would rather sell you the machines than contract'.

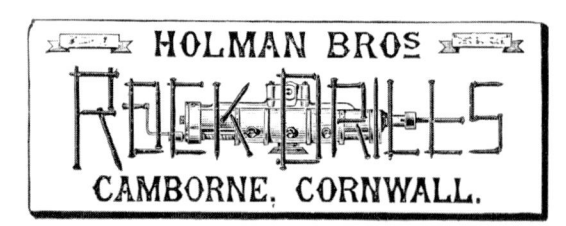

A rather curious Holman collophon used in the early 1900s.

As mining agents, Holmans still recruited men for foreign mining fields. 'Wanted. A good rock drill man for six months. Knowledge of Spanish and Holman Drills essential' was a typical advert. Edward Rowe who went to Margrahamore in Ireland to install 'a Holman steam drill' was asked to stay as Captain and received a presentation on leaving. Holmans own mineral dressing plant and pneumatic stamps established a world wide reputation hence, 'Wanted for Bolivia. First Class Tin Dresser. Apply with references to Holman Bros. Camborne'.

Business was so brisk that new machine tools were regularly added, as in

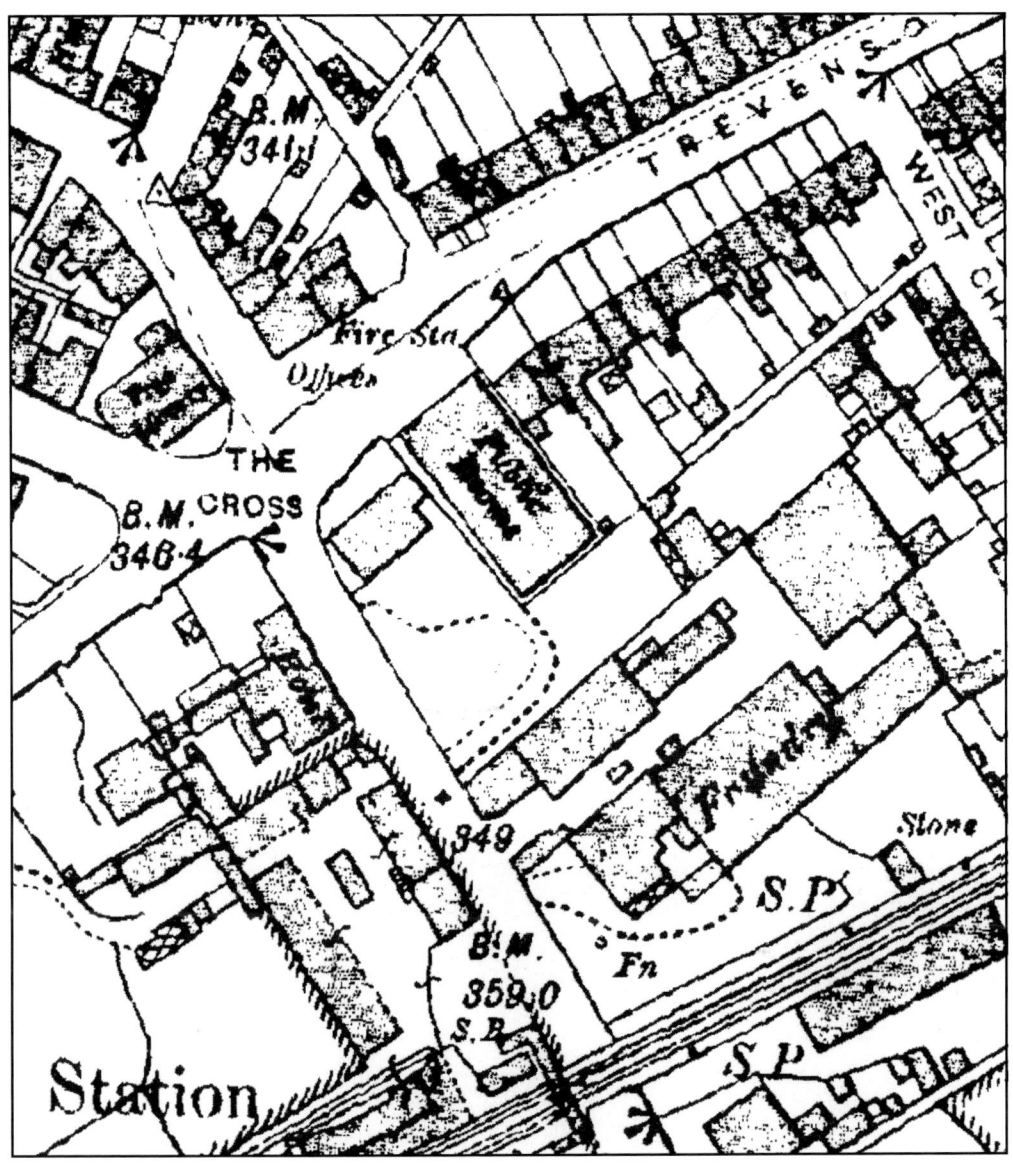

1908 Ordnance Survey map showing the early No.3 Works and the Public Rooms.

December 1903 when Thomas Ward of Sheffield sold off the remaining machinery at Harvey's Foundry. Holmans bought a variety of centre lathes including, for £200, a big hexagon turret lathe by Alfred Herbert of Coventry. An 8-ton ladle for the foundry cost £15 and a wrought iron jib crane slightly more, with a steam hammer being procured for £92 'prompt cash'. A punching and angle cutting machine cost them £40 and an air receiver just £18.

The closure of Harvey's Foundry had benefited Holmans, who employed many of the redundant young apprentices and skilled men. William Rodda came to Cornwall from Cheshire in 1890 and moved to Camborne three years later. After a spell at Climax he joined Holmans in 1897, becoming one of the original team who set up the new No. 3 Rock Drill Works. Joseph Tromans began his career as a chain maker at Cradley Heath. When the Harvey Foundry declined he brought his family to Camborne in 1884, at the invitation of James Miners Holman, providing he could have a house. He spent the next thirty years in the smiths shop while his son Thomas, aged fifteen, who had worked alongside him at Harvey's, spent the next sixty years as faggoter, smith and tool sharpener. The Tromans family produced many good musicians including Thomas, who was a founder of the famous Holman Male Voice Choir. Thomas's son Percy would eventually become manager of the No. 1 Works.

There were still orders for big compressors and another one was being made for Carn Brea and Tincroft when a young apprentice pattern maker joined the company on 30 April 1904, the

> …patterns were being made and some of the castings were well advanced. There were no draughtsmen in my early days. The compressor was marked out in pencil on a big whitened board which was practically the length of the pattern shop. Then later drawings were made from the finished parts. The compressor was too big for us to assemble at the works. When Mr Plummer joined, the old system got a shaking. He introduced the idea of preparing drawings BEFORE the manufacture of engines and so on. Before then, the practice had been the other way round. Plummer came to the works in a silk hat. Never missed.

In December 1906 East Pool contracted with Holmans for a new compressor to be delivered in twelve months but 'owing to the expeditious way the company always carries out its contracts, the compressor has not only been delivered and placed in position (June 3rd 1907) but is practically ready to start work'.

No big Cornish pumping engines had been built since the closure of Harvey's of Hayle but Holman Brothers had the distinction of making the very last of them. In January 1903 the Dorothea Slate Quarry in North Wales decided to install a large pumping engine to cope with a serious inrush of water. Nicholas Trestrail was asked to inspect two old 70-inch engines lying in Cornwall but neither was

A remarkable image of a boiler which is presumanbly being backed out of No.1 Works. In the background the office block has scaffolding around it, suggesting that it was still being built; this dates the photo at 1906. On the right, the building is single storey; shortly after this it became two stories high, part of the No.1 Heavy Machine Shop.

suitable. Accordingly he was asked to design a new 68-inch capable of working 18-inch pump work. Holmans' tender for £1,925 was accepted while the Redruth Foundry was contracted to cast the heavy pitwork. Nicholas Trestrail's 'Plan of Pitwork' was received at Dorothea on 25th April 1904 and the 'Big Shaft' was prepared to receive the new engine. The 82 tons of pitwork, apart from two long lengths of rising main and a windbore sent on by rail, was shipped aboard the Bideford ketch *Iron King* which sailed from Hayle on 3rd September 1904, arriving at Caernarfon five days later. The bob was despatched on a G.W.R. flat car from the yard of No. 1 on 14th October 1914. Six days later the quarry manager noted that the

Big Beam of the new engine arrived at Nantille Station from Messrs Holman Bros. of Camborne Cornwall, through Mr Trestrail. Weight 22 tons 12 cwt. A traction engine belonging to Mr Robert Roberts of Abergele arrived to haul the heavy parts to the quarry.

During the following week the rest of the engine arrived from Camborne which was duly noted.

October 21st Unloading Big beam at station'

October 25th Cylinder bottom, weighing 5 tons 11 cwts. Cylinder Cover weighing 3 tons 10 cwts and Piston and Rings weighing 2 tons 198 cwt 3 qtrs, were hauled from station to quarry.

October 26th Big beam was hauled to Quarry.

October 28th Cylinder arrived at station weighing 16 tons 17 cwts 1 qtr.

October 29th Hauling cylinder to the quarry.

The top of the 68-inch cylinder and the valve chest of the Dorothea engine. Photo: Dave Sallery, www.penmorfa.com.

Evidently the rough weather considerably delayed operations and not until 4 April 1905 was it recorded that 'Thos. Rowe arrived from Cornwall to superintend the erection of the engine'. Next day, having been joined by his son Percy, they began 'Rising the Bedplate of Big beam to the top of the front wall'. On the 17th they were engaged 'Lifting Big Beam to its place'; ten days later were busy 'lifting and fixing the cylinder' and on 15 June 1905 'Thos Rowe and his son left for home after finish erecting the engine'.

There was another long delay but on 5 February 1905 'George Perry 'Pitman' arrived at the quarry from Cornwall to superintend the fixing of pumps'. Manoeuvring and adjusting the heavy cast iron work in the cramped shaft was a laborious task, though Perry used a capstan engine from a Victorian battleship lately scrapped by Thos. Ward of Sheffield. On 8 May 1906 Thomas Rowe was back 'from Cornwall to start the engine' and four days later the 'new engine moved for the first time, in order to take away some props that were holding the rising main'. On 14th May the 'new engine and pumps starting to work for the first time' and Nicholas Trestrail soon arrived to check the work. Adjustments were needed but on the 19th 'Thomas Rowe left for Cornwall and W. H. Curnow arrived in order to learn our men how to work the new engine'. In a month he and George Perry left for Cornwall and the 'new engine began to pump on 19th July 1906' and would stay in service until replaced by electric pumps fifty years later.

Next year Holmans was asked to build another large Cornish pumping engine, an 80-inch, again designed by Nicholas Trestrail. This was for the re-

The 80-inch South Phoenix engine. After the mine closed in 1914 the engine was preserved in grease; it was scrapped in 1931.

working of South Wheal Phoenix near Liskeard. The bob was fabricated from rolled steel plates by Holmans but the enormous cylinder was subcontracted to Worsley Mesnes Foundry of Wigan.

Holmans were still called upon to repair or refurbish some of the few Cornish engines left at work around Camborne. They supplied a new bottom for the cylinder of the 80-inch pumping engine on Lyle's Shaft at the Basset Mines, after the main rod broke and the bob came 'in doors'. One notable job came after the old Harvey bob of Goold's eighty at Wheal Grenville suddenly broke in half at the gudgeons late on a September night in 1906. Six tons of cast iron crashed down on the steps where the engineman and his mate had been sitting just a few minutes before, but was stopped from falling into the shaft by the pump rod. Immediately there were rumours that the engine was wrecked and the mine flooding but on arrival, the correspondent of the *Mining Journal* was confronted by Captain Negus climbing out of the big cylinder and accepted his offer to inspect the damage, which was not that devastating.

Marshall's engine kept the water down while Captain Negus summoned Holmans' engineers, who only a month before had been underground to repair a cracked balance bob in the same shaft. The new

The new half beam being raised into place at Goold's engine house, Wheal Grenville.

The duplex horizontal winding engine built for South Crofty Mine. These were also built for Botallack and South Phoenix mines, all parts of the Cornish Consolidated Tin Mines empire.

bob for the eighty was 38 tons to the old one's 23 and as a precaution horns were cast on the underside of both ends to anchor steel bridles which would brace a kingpost should the same accident happen again. So urgent was the work that instead of allowing the mass of iron to cool for two or three days the foundrymen made thick gloves from Hessian bags to enable them to free the bob from the casting mould.

The bob was raised by a 'willing gang of Grenville miners' who worked day and night under Captain Negus, and were rewarded for their efforts by Peter Watson, the chairman. On 7 November 1906 the directors, who found nobody to blame for the accident, heard that 'the new bob which has been cast with greater weight and improvements made thereto - by Holman Brothers Ltd. - has been placed in position. The engine was set to work on Saturday last, the 3rd instn. and has continued to work smoothly and satisfactorily during the last week.'

Holman compressor being pulled across the veldt by oxen, around 1906.

The Grenville bob was a remarkable success for Holmans who already had a new winder in their shops for South Crofty, at that time the newest and ultimately the last of all the Cornish tin mines. A big Cornish pumping engine had been set up in 1903 on the new Robinson's Shaft, sunk like Williams at Dolcoath to avoid a maze of old workings and crooked shafts. By 1906 Crofty was controlled by Cornish Consolidated Tin Mines Limited, a speculative and labyrinthine conglomerate of companies relying upon a high tin price mainly to rework long closed tin mines. Surprisingly, South Crofty had defected to Climax after buying Little Vixens and the big Imperials, beginning a tradition that would last fifty years. However the mine came to Holmans for a new 400 HP steam winder, a 22-inch twin cylinder direct acting horizontal, with double beat drop valves and a 4-foot stroke, driving one fast drum which was permanently keyed to shafting and a second drum that could be clutched in or out. The winder was steamed by two Lancashire boilers and well equipped with a Green's 'Economiser', two Thompson Feed pumps and a Wheeler condenser. During the spring of 1907 foundations for the engine were laid, being 'concrete throughout and not a single one of those granite blocks which formerly gave the old 'loadings' the jumps'.

The engineman recorded, 'Started with single cage fixed drum ...8th April 1907'. By July, the winder was ready to hoist double decker cages on 1⅛" steel wire at 1,000 feet a minute via a new and massive timber headgear mounted on concrete piers. Crofty had also ignored Holmans and erected a big 'Reidler' air compressor together with two steel Lancashire boilers, air receiver, cooling tower and reservoir and a 100 steel chimney. Robinson's winder, as it quickly became known, would, with various modifications, work uninterrupted for the next fifty years.

In January 1908 Holman Brothers opened their new showroom in Wesley Street which attracted a great deal of attention, especially the photograph of 'three huge elephants in the track of a boiler.... two apparently pushing it. The steep mountainside is admirably depicted while on the other side a band of swarthy Orientals are seeing Fair Play'. There were also displays of drills and another large photograph showing natives hauling a boiler with a Cornishman supervising the operation.

Elephant-powered transport in India.

Holmans also had an excellent display at that summer's exhibition at Olympia and brought it all home for the Polytechnic Exhibition held in Camborne Market House during the first week of September. The continuing success of Holman Brothers was sadly interrupted by the death of John Henry Holman at the early age of fifty five at *Tregenna* on 4th November 1908.

Johannesburg stores in 1906.

Chapter Four

The Gold Reefs

'I have great pleasure in testifying that during the month of January 1907, at the Kleinfontein Deep Ltd... 131 feet were sunk in 13 drilling rounds.

This I believe constitutes a world record and was performed with your 3¼" 'Holman Drills'. This record should establish an undoubted efficiency which you claim for your drills.'
Benoni Transvaal.

If the discovery at a Boer farmstead in 1887 had been tin or copper, history might have been different, but the strike was gold, the largest deposit in the world and concentrated along a range of low scrubby hills, the 'Ridge of White Water', the Witwatersrand. Henceforth, gold mining and South Africa were indivisible and this was to shape the lives of generations of Cornish miners and engineers.

Imposing granite villas around Camborne and Redruth still echo the days of 'South Africa the Golden', with names like 'Crown Reef, 'Ferreira Deep' and 'Randfontein' still discernible on marble plaques or faded glass fan lights. After the great tin slump of 1895 the Witwatersrand became the El Dorado for every Cornish miner tired of long shifts and poor pay. Baggage and trunks bearing the labels 'Capetown' and 'Johannesburg ' became familiar sights on railway platforms from Penzance to Liskeard, while there was the Friday night ritual of seeing off husbands, fathers and brothers who crowded into carriages whose windows bore the sticker 'Southampton'. Union Castle became to the Witwatersrand what Cunard was to Michigan and P & O was to Kalgoorlie.

Diamonds at Kimberley and gold on the Witwatersrand drew thousands of Cornishmen into South Africa, miners, engineers, artisans and small tradesmen. There was big money to be made; the prudent or the lucky came home and bought themselves a coal yard, grocery store or public house. More transitory and flamboyant was the spectacle on Saturday nights when the 'Africa men resplendent in digger hats and gold watch chains', pushed their way through the doors of Tabb's Hotel in Redruth or the Commercial at Camborne, often ending the night sprawled in the gutter or being kicked off the last tramcars.

Travellers on a Union Castle steamer on their way to South Africa.

The Cornish survived all the political upheavals. Cecil Rhodes might have smuggled 200 rifles and a million rounds of ammunition in oil drums to the discontented 'Uitlanders' in Johannesburg but the Cornish had come to dig gold, not join a revolution against their Boer landlords. They decamped in railway cars carrying the derisive slogans 'For Cornish and kaffirs only' but many did serve in the Matabele Wars or later when the Boer War itself broke out, though scores came home where Cornish mining had revived, aided by the introduction of electricity.

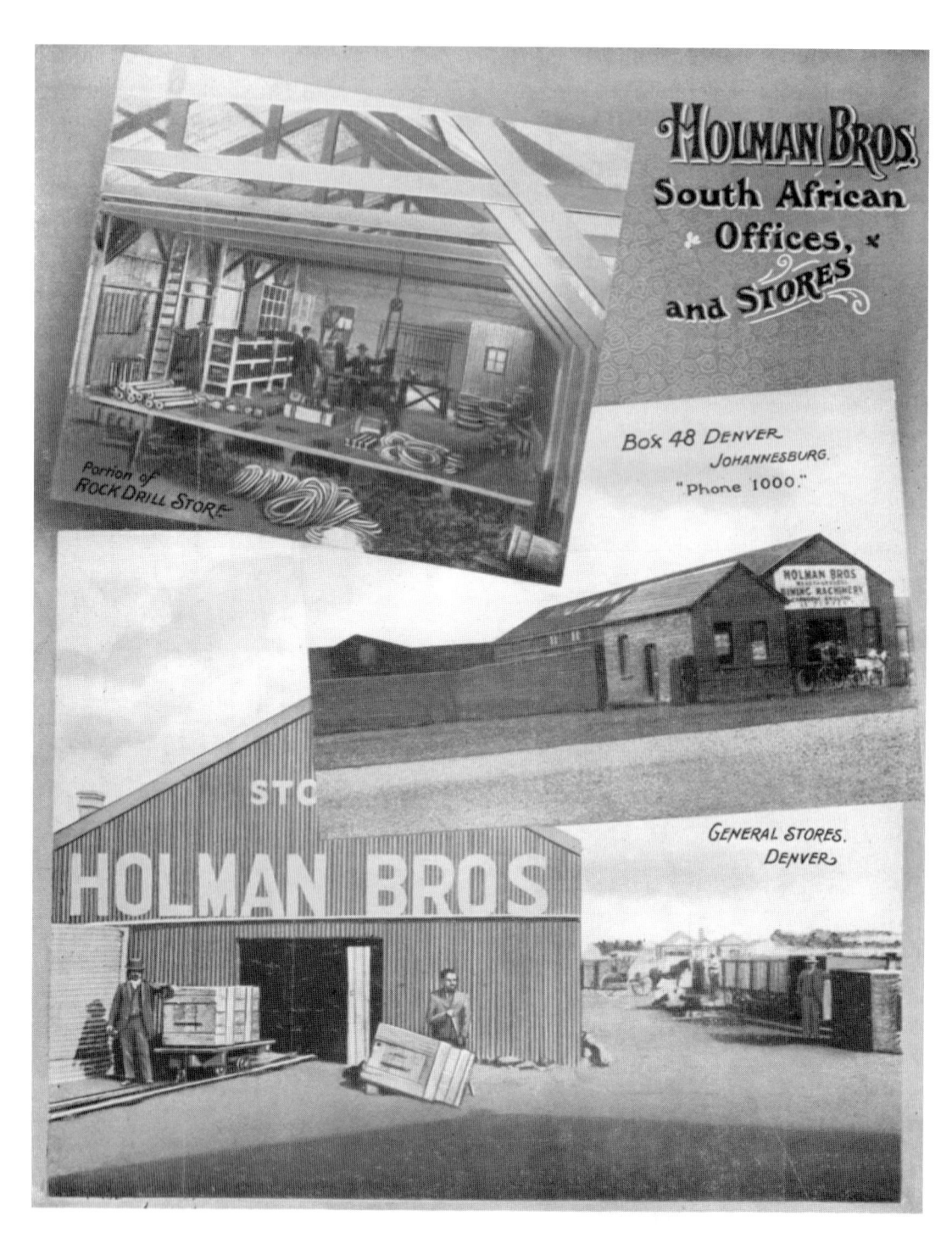

Holman Brothers South African catalogue for 1906.

When the war ended in 1902 the country was in a desperate state, especially after three years of highly effective guerrilla war which had done somewhat less damage than the heavy-handed counter-measures of the British military. The massive reconstruction of the country relied upon the wealth of the Witwatersrand but the mines were in a parlous state when the Boer field armies quit the Rand in the spring of 1900. Only Geldenhuis, which was owned by a neutral French company, had stayed at work and when the British army was advancing, 'Oom Paul' - Paul Kruger, the Boer president - had threatened to blow up every mine on the Rand, an Armageddon more akin to the Old Testament than the practical application of dynamite to 43 separate mines.

Yet the gold mines were recognised as the salvation of South Africa but they had to be not only refurbished but developed at a pace never before considered. The answer, like that in Cornwall thirty years before, was mechanisation, haulage, winding, stamping and especially machine drilling. The Rand became a Mecca for American, German and English engineering companies, all eager to cater for

Record-breaking drilling crew at Kleinfontein Mine, 1908.

The record-breaking drilling team at Brakpan Mine.

a market where they stood to make millions. Holmans themselves were ready for such a boom, especially as the Cornishmen on the Rand were certainly going to make sure the new machines came from home, though their own tin mines were dwarfed by those on the Rand. By 1909 there were 2,389 rock drills boring the gold reefs while 2,000,000 tons of rock were being hoisted to surface. Small wonder that John Holman had hoped one day that they would be 'able to exploit Dolcoath on the lines obtained in South Africa', though the deepest gold shaft, the Catlin at Jumpers Gold Mine, was already 1,000 feet deeper than Williams Shaft.

He had visited the Rand in 1904 accompanied by his wife and two Holmans men, Frank Luke and W. J. Trythall, who were to open a stores and workshop in Johannesburg. Soon to follow was young Arthur Reynolds who was from an old St Agnes mining family and the son of the manager of Charlestown Foundry. Tired of a 'sedentary' clerk's job in No. 3, he chose to try his fortune on the Rand.

So successful was he that twenty years later he returned to establish a thriving coal business, supplied by his own steam colliers sailing from Portreath. His young cousin John Reynolds, spent some time at No. 3 before going over to work under Mr Miller at No. 1 until, in 1911, he was transferred to the short lived New York office.

Among Holmans agents were Howard Farrar & Co. and the ebullient William Hosken of Hayle or 'B'Losken' to his fellow Cornishmen. He was a pioneer 'Uitlander' who had sat on the 'Reform Committee' during the Jamieson Raid back in 1895. He always preferred the turbulent 'politicking' of the Rand to mere storekeeping and a legacy of this was the bundle of dynamite found buried under the porch of his home 'Trelawney' in Jeppe's Town.

James Holman's trip to the Rand was very successful and by early 1905 'the most satisfactory sounds in the evenings are those of the hammers of Messrs. Holman Bros. No. 3 factory where they are working overtime on a good order from South Africa.' The *African World* also carried a long article on the big air compressor at Dolcoath, which appears to be a vigorous reply to the publicity campaign waged by Climax whose adverts easily outshone those of Holmans in the *Cornish Post* or *Mining Journal*.

Yet Climax and all the other makers of mining machinery faced enormous technological challenges out on the Rand. The gold mines demanded speed and

Early method of allaying dust - a simple water spray attached to the rock drill.

79

mechanisation and there was no place for the old traditions which encumbered the Cornish mines, though both the Cornish miners and engineers once in South Africa never hesitated to embrace all the new mining methods.

Rock drill repair shop in South Africa in the early 1900s.

There was a price tag attached to this amazing technological development and that was 'phthisis', the destruction of a miner's lungs by the dust of the gold stope, hard silicate dust so fine that it could seize the bearings of a theodolite. Drilling, blasting and shovelling were all done 'dry' amidst clouds of dust. The result was catastrophic, especially among the Cornish miners and soon there was 'scarcely a village in the Duchy which has not seen a victim of this disease'. Both Holman and Climax produced patent dust allaying water sprays but the Cornish miners were notoriously indifferent. It was to take decades before regulation and development of the rock drill were to reduce the dreadful toll of phthisis.

The great quest was for a small machine to open up the narrow gold stopes as

Boiler Works manager 93
Warren, T. 189
Watson, Peter 71
Weekes, R. 182
Wesley Street offices moved to Rosewarne
    158
West Ham United FC 300
West of England Steam Engine Society 322
Wethered, Oliver 50
Wheal Grenville
  air compressor for 89
  Holman rock drills at 45
  new beam for Goold's engine 70
  removal of 90-inch engine from 111
Wheal Kitty
  new bob for 90
Wheal Seton
  Holman rock drills at 45
Wheal Vor 15
White, Geoff 319
Whitehead, Harry 196
White, Major Richard 60
Whittle, Sir Frank 240
Williams, D. 177, 187, 189
Williams, Ernie 161
Williams, J. M. 189
Williams, Michael Henry
  cuts first sod of Williams' Shaft 49
Williams, R. 181
Williams' Shaft 49, 53
Williams, Victor 116
Wilson, C. 281
Wilton, B. 189
Witwatersrand 74, 77
Wombat
  hand-powered drilling machine 106
Woodward, S. 181
World War II 133
Worsley Mesnes Foundry
  casts South Phoenix engine cylinder 70
Yale trucks
  games with in works 249
Yekaterinburg 106
Yeo, William 86
Young, Arthur 321, 323
Zitair compressors 308

Symons, John 185
Szecheun (Sichuan)-Yunnan Railway
  orders for 131
T.100 turbo compressor 241
Tabb's Hotel, Redruth 75
Takoradi Office 116
Tangye, George Bertram 19, 22
Tasker, Andy 186
Teague, Captain 38
Teague, Captain William 27
  orders Cornish rock drills for Carn Brea
    Mine 34
Teague, William junior
  developes Cornish Boy rock drill 34
Tehidy Saw Mills 22
Tennis Club 109, 179
Thames Iron Works 282
Thames Iron Works FC 300
Thomas, Arthur 106
  reports deaths of Russian royal family 107
Thomas, Captain Charles 45
Thomas, Captain Josiah 27, 35, 45
  and Barrow rock drill 28
  death of 60
Thomas, C. V. 129, 198
Thomas, George 189
Thomas, Gladys Fredericka
  marries John Leonard Holman 85
Thomas, Herbert 41
Thomas, R. Arthur 49, 51, 113
Thomas, William 210
Thorneycroft Lorry 124
Tincroft Mine 27
  fire at compressor house 49
Tiverton Heathcoat Cricket Club 191
TM 60 compressor 150
Tolgus Shaft Company 113
Tolgus Tunnel 110, 113
Tolgus Tunnel Company 109
Torbay, Alf 185
Toye, J. C.
  sells steam engine to Stephens 195
Tractair 160
Traction engine
  accident with 21
  acquired by Holman Brothers 20

Trebilcock family 303
Tregonning, A. H. 177
Tremelling, Alfred 197
Treskillard tin stream
  steam engine for Stephens 196
Trestrail, Nicholas 49, 56, 63, 67, 69
  designs South Phoenix pumping engine 70
  installing Dorothea engine 67
Trevithick Day parade 324
Trevithick Memorial Committee 129
Trevithick, Richard 14
  steam rotary borer 23
Trevithick Scholarship 129
Trevithick Society 131, 157, 240, 313, 319,
  321
Trewin, W. C. 102
Tromans, Joseph 66
Tromans, Percy 66, 130
Tromans, Thomas 66
Troon Rugby Club 187
Trounson, Jack 131
Trythall, W. J. 78
T Type compressor 124
Tucker, F. 177
Tuckingmill Foundry 64
  closure of 85
tunnelling pump 106
Ulathorne & Co. 30
Union Castle Line 74
Vane type air motor 124
Vanstone, Leslie 185
Veale, John 122
Veal, John 22
Vee engines 129
Verran, E. 177
Village Deep Gold Mine 83
Vivian, C. F. E. 116
Vivian, 'Farmer' 112
Vivian, J. H. 116
Wade, Jethro Thomas 269
  leaves Broom & Wade 272
Wakeham, Gerald 189
Wakeham, H. 189
Warmington, D. 189
Warren, John 93, 197

Salmet, Henri 93
Sanders, Nick 314
Saundry, Nicky 112
Sawle, John 322
Scapa Flow
  salvage of German warships at 124
Scott, William Harding 281
Selwood, C. 189
Selwood, Foster 189
Selwood, Pat 189
Sergeant, Henry Clark 25
Serving the World film 151
Sherlow, Miss M. 181
Siebe, Augustus 305
Silver (Cornish) 82
Silver Bullet 150
  technical problems with 154
Silver Feather 257
Silver Three 160, 257
Simons, Alf 112
Simpson, George Darlington 210, 215, 216, 218
  resolution to have removed as director of Climax 214
Singer brothers
  rock drills 23
Singer, Isaac 23
Skink. *See* Polsten
SL9 rock drill 141
SL.14 drifter 250
SL. 280 drifter 153
Smith, Arthur Herbert 57
  death of 149
Smith, Christopher Broom
  joins Broom & Wade 274
Smitheram, Leonard 180
Solomon, A. 187
"Solomon of Tuckingmill", the 44
Someiller, Germain 26
South African Chamber of Mines
  prize for best rock drill 82
South African Mining Journal
  sponsors drilling contest 81
South Crofty Cricket Club 182
South Crofty Iron Works 35
South Crofty Mine 207
  buys Climax rock drills 72
  Robinson's Shaft 72
  steam winder for 72
  winding engine for New Cooks Shaft 89
South Phoenix Mines 96
South Roskear Mine 29
South Wheal Crofty 202
South Wheal Phoenix 70
Soviet Union
  sales to 128
Sports and Social Club 168
St Allen 12
St Austell Brewery
  loans to Sports Club 182
Stephens, Dorothy Annie 203
Stephens family
  sales of Climax shares by 229
Stephens, Mary Ann 193
Stephens, Muriel Audrey 203
Stephens, Richard 192
  death of 204
Stephens, R. & Son 35, 192, 194
Stephens, Thomas 192, 193
Stephens, William Charles 35, 115, 192, 193, 196, 201, 210, 218
  appointed MD of Climax 212
  becomes joint MD of Climax 222
  death of 226
  elected Vice-President of the Royal Cornwall Polytechnic Society 210
  goes to South Africa 223
  overcome by funes at South Wheal Crofty 196
  reduction of salary and directors' fee 225
  resigns from Climax 225
Stephens, W. R. 181
St Gotthard Tunnel 38
St Just foundry 16
Strauss, Arthur 50
Stream Lined (SL) drills
  introduction of 117
stretcher bar hoist 88
Strike, J. 177
Superhaul 117
SV 40 rock drill 255
Symons, A. 181

Rand Drill Company 25
Randfontein Estates
  Climax record at 225
Reavell, Captain Kingsley
  joins board of Reveall & Co 293
Reavell & Co 281
  Quadruplex compressor 282
Reavell & Company Ltd
  incorporation of 282
Reavell, Lieutenant-Colonel K.
  forms Home Guard unit 295
Reavell-Mossay Pneumatic Tool Company 293
Reavell, William 281, 286, 290
  appointed Knight Bachelor 295
  becomes Chairman and Managing Director 297
  first President of IME 293
Red River 12
Reed, Mary 13
Regional Development Agency 315
Reveall & Co
  acquired British Empire rights to patent by Aerzener Company 292
  aqcuires land for works extension 293
  Axial type compressor 289
  compressor for Diesel engine 288
  Gray-Wimperis gyroscopic bomb-sights 290
  portable petrol-engine compressors 293
  production of compressors 286
  Ranelagh Works 283
  Ranelagh Works Athletic Association 298
  single-stage, single-acting, vertical compressor 292
  Sports club 298
  tandem two-stage quadruplex compressor 287
  three-stage quadruplex compressor 288
  V class compressor 288
  work during Great War 290
  work during Second World War 295
  works enlarged 289
Reynolds, Antony 186
Reynolds, Arthur 78
Reynolds, J. H. 157

Reynolds, John 79
Rhodes, Cecil 75
Richards, M. 189
Richards, Mr
  rock drill design 198
Richards, Ray 186
Richard Trevithick
  unveiling of statue in Camborne 129
Rickard, C. D. 182
Rickard, Kingsley 321
Rio Tinto 31, 38
Rio Tinto rock drill 41, 198
Ritchie, James
  appointed joint Managing Director of Holman Group 252
Road Breaker 124
Road Ripper 124
Roberts, J. 177
Roberts, W. J. 181
Robins, G. 189
Robinson Deep Gold Mine 83
Robinson's Shaft 207
Rock Drill Hospital 29, 35, 195, 197
Rodda, William 66
Rodda, William S. 157
Rogers, Mo 186
Rosewarne House 87
  becomes Gladys Holman Home for Spastics 256
Roseworthy Hammer Mills 86
Roskear Fuse Works 60, 105
Rotair 302
  rotary screw compressors 241
Rowe, Anne 12
Rowe, Edward 64
Rowe, Percy 69
Rowe, Thomas 69
Royal Cornwall Polytechnic Society 209
  medal and diploma won 124
  rock drill contest 35
R. Stephens & Son 29
Rugby team 109
Rule, W. David 116, 161
  death of 242
Rundle, Sir Alexander 201
Rutherford, Andrew 216

Moreing, W. C. 94
Morgans, Charles
  designer of Williams' Shaft winder 53
Morgan, Thomas 60
Mountain Queen Mine, Western Australia
  pneumatic stamps at 85
Munitions XI football team 184
Mysore Mine 200
Negus, Captain 71
Nettle, Mr 193
Nettle, Sydney 122
New Dolcoath 113
  headframe for Roskear Shaft 114
"New" Factory
  for Polsten production 146
New Sump Shaft (Dolcoath Mine) 29
No. 3 Rock Drill Works 66
No.3 Rock Drill Works
  building on site of 318
Noble, Leila
  visits bottom of New Sump Shaft 29
North Roskear Mine 47
Oates, D. L. 181
Oats, Major Francis 50, 60
  death of 109
Odger, A. 177
OEM Fabrications 315
Oerlikon gun
  manufacture of 145
Okuno, Frank Trevithick 325
Old Sump Shaft (Dolcoath Mine) 28
Old Wheal Agar Engine Shaft 109
Opie, Mary Anna
  marries Willaim Stephens 203
O'Reilly, P. J. 181
Pachuca Mines, Mexico
  equipment for 89
Palmer, D. A.
  joins board of Reveall & Co 297
Paris Exposition, 1900 54
Park Iron Pit 28
Parnall, James 103
Parnell, Reg 189
Pascoe, H. 177
Paul, Hugh
  death of 297
  joins Reveall board 289
Paull, Captain Josiah 89, 111
Paull, Josiah 207
Paul, William 289
Paul, W. J. 46
Pearce, Gilbert 60
Peat, Sir William Barclay 212
Penberthy, John 197
Penhaligon, David 245, 306
Penhall, Captain 49
Penhall, T. 44
Perry, George 69
Peru 15
Peters, G. D. 295
Phillips, Charles Trezona 86
Phillips, W. D. 177
Phthisis 80
Pidwell, Thomas Garfield 102
Pitt, Neil 186
Plummer, George
  Holman Brothers' chief designer 63
Pneumatic tools
  development of 125
Pneumatic tool shop
  moves from Public Rooms 150
Polsten antiaircraft gun
  manufacture of 145
  problems with 147
Pooley, F. 177
Power Jets (R & D) 240
Powning, Jack 86
Prideaux. Len 189
Prince of Wales and Duke of York
  visit to Camborne of 121
Princess Royal
  visit to Camborne works 313
Pryor, Miss E. 181
Puffing Devil 14, 313, 319
  centenary of 64
Quaife Engineering 315
Queen Mother
  visit to Holman Brothers 257
Queen's Award to Industry 256
Quertier, R. L.
  joins Reavell board 293

Gundry, L. 181
Halfyard, David 191
Haly, C. 181
Harris, Andrew
  saved William Stephens 196
Harris, H. D. 177
Harris, William 122
Hartley, H. A.
  elected chairman of Reveall & Co board 297
  Reveall works manager 286
Harvey, Ann 14
Harvey, E. 181
Harvey, Frank 50, 52, 60
Harvey, H. 44
Harvey, Harry 122
Harvey, Lieutenant Cyril 103
Harvey's Foundry
  closure of 65
Harvey, W. F. J. 22
Hashing, G. 44
Hashing, J. 44
Hathorn & Co. 31, 36, 37
Hathorn & Davey engineers 56
Haupt, General Herman 24, 27
Havelock
  home of Richard Stephens 203
Hayman, Frank 189
HD3 hammer drills 110
Heal-Shaw Beecham air motor 117
Heavy cradle Drifter (Climax) 222
Heavy Machine Shop
  alterations to 254
  construction of 99
Hills, Arnold 299
  retired from Reveall board 289
Hills Arnold F. 282
Hirnant rock drill 199
Hitchens, Lieutenant-Commander Robert 141
Hocking, C. 181
Hocking, J. 177, 181
Hodge, C. 177
Hodge, James (Jim) 240, 302
Holman apprentices
  pay rates 116, 121
Holman, Arthur Trevena. *See* Holman, Treve

Holman Bowling Club 189
Holman Brothers
  becomes private limited company 85
  London office opens 57
  150th anniversary 155
Holman Brothers Ltd
  quoted on London Stock Exchange 241
Holman Canteen 150
Holman Concert Hall 150
Holman Cricket Club 182
Holman, Dorothy 94
Holman, Emanuel 13
Holman Football Club 184
Holman, G. 182
Holman, Grace 13
Holman Group
  formation of 236
Holman Group Magazine 151
Holman Iberica, S. A.
  new factory built 243
Holman, Jack 93
  sent to France as Lieutenant 96
  shot by sniper 98
Holman, James 13, 16
Holman, James F. 253
  appointed Chairman of Holman Group 250
  becomes director Holman Group 242
Holman, James Henry 18
Holman, James Miners 18, 20, 50, 56, 60, 99, 101, 105, 109, 129
  acquires Rosewarne House 87
  attends Imperial Exhibition 84
  becomes director of Dolcoath Mine 52
  becomes J.P. 59
  chairman of East Pool Mine 94
  death of 130
  visits the Rand 79
Holman, John 13, 16, 18, 20, 21, 22
  death of 41
  foundry next to Wesleyan Chapel 17
  retires 31
Holman, John Henry 20, 78
  death of 73
  director of Dolcoath 49
Holman, John Leonard
  becomes director 85

335

Dower, R. 177
Downing, Handcock, Middleton & Lewis
 solicitors 210
Dryductor 302
Duke of Kent 129
Dunkerton, E. C.
 first chairman of BCAS 127
Dunstan, R. 181
Dunstan, S. J. 181
Dunstanville, Lord De 13
Dust Allayer 205
Dust allaying water sprays 80
Dustuctor 155
East Grenville Engine Shaft
 deepenng of 46
Eastman, Jon 322
East Pool Mine 31, 38
 air compressor for 89
 collapse of workings 109
 new whim for 43
Eclipse rock drill 35, 197
 at East Pool Mine 31
 at Wheal Cock 31
 wins Silver Medal 31
Eclipse valve 25
Elkington, T. R. 281, 282
El Oro Mine, Mexico 106
Endsleigh House
 home of William Stephens 203
English Partnerships 315
Eustice, George
 death of 63
Evans, Joseph Foundry 192
Ewing, Alfred
 appointed joint MD of Climax 222
Ewing, Ralph 222, 236
 Chairman of Climax 232
 death of 242
Excelsior rock drill 35
Falmouth Docks 31
Favill, H. L. V. 188
Feirrera Deep Gold Mine 83
Fiddes, N. G. 177
Finch, H. K.
 Reveall director 286
Fitch, Peter 308

Floyd, J. 177
Fluid Logic Control System 239
Ford, J. 177
Fowle-Burleigh patents 25
Fowle, J. J. 24
Fraser & Chalmers Ltd 116
F.T.A.C. (Italian company) 125
Garnder Denver Inc. 318
Garstin, Norman 91
Garth House (Welsh Office) 116
Gaskell, Charles 281
Gay Nineties Club 177
G. D. Peters & Company Ltd.
 Reavell & Co agreement with 294
General Motors
 BroomWade agreement with 271
Gifford, Captain William 137, 138
Gilbert, Captain John 200
Gilbert, Dick 118, 123, 130
Gilbert, Miss G. 181
Glasson, Henry 208
Glenfield & Kennedy Ltd 283
Godolphin family 303
Godolphin, Sid 307
 union convenor 306
Goodyear, James Wallace 239
Goodyear Pump 239
Gooninis Mine
 hoist for 58
Great Dowgas Mine 96
Great Flat Lode 56
Great Stope Drill Contest 83
 won by Holman Brothers 84
Great War
 alterations to Wesley Street works 99
 Holman workers volunteering for 101
 recruitment of girls and young women
  during 103
Green, George 197
Greensplat pumping engine 239
Grenville United Mines 22
Grootvlei Mine
 World Underground Development Record
  broken at 153
Grose, Captain Samuel 14
Gundry, E. T. E. 237

acquired by Invensys PLC 313
acquires AQUAP designation 311
closure of 313
rebranding name 308
sold to Alchemy Partners LLP 313
sold to Gardner Denver Inc 317
CompAir Industrial 302
CompAir Industrial Products 317
CompAir International 302
CompAir Maxam 302
CompAir Power Tools 312
CompAir Tools & Equipment 312
Consolidated Brake Company 295
Copperhouse Foundry 15
Corliss valves 56
Cornish boiler 14
Cornish Boy rock drill 34
Cornish Consolidated Tin Mines Limited 72
collapse of 96
Cornish Electric Power Company 89, 110
Cornish Engineers
150th anniversary book 157
Cornish Engines Preservation Committee
formation of 131
Cornish Engines Preservation Society 319
Cornish Industries Fair 157
Cornish rock drill 35, 38, 40, 198
in contest at Dolcoath Mine 33
Cornwall Boiler Company 47
Cornwall Boiler Works 93
The Cornwall Drop Stamping Company 47
Cornwall Drop Stamping XI football team 184
Cornwall House
staff accommodation in Takoradi 116
Cornwall Railway Company 193
Cornwall Works, Birmingham 19
Cory, Wearne 177, 182
Couch, John 23
Cowling, T. G. 181
Cox, Garstin 44, 90, 109
death of 130
Cox, William 90
CPR Regeneration Company 315
Crabbe, Samuel 194
Cradle for supporting rock drills 206

Cricket club 109
Crocker, Alfred J. 212, 214
Crofty Cricket Club 183
Crofty/Holman Cricket Club 183, 184
Crown Deep Gold Mine 83
Holman sinking record at 116
Crumbs from Cornwall (booklet) 221
Crystal Palace Exhibition 1890 199
Crystal Palace mining exhibition 40
Curnow, W. H. 69
Curry, Nicholas (Nicky) 22, 99, 116, 122, 157, 159
Curzon, C. W. 218

Daddow, W. A. 181
Dagenham Girl Pipers 157
Darlington, Colonel 197
Davey, Henry 56
David Ball Construction 320
Davies, N. Baldwin 236
Deepdale Engineering Co. 320
de Navarro, Jose F. 24
Desborough
closure of compressor plant at 305
Diamond Rock Boring Co. 30
Dispute between Holmans and McCulloch 39
Doering, Frederick
tests drill at Tincroft Mine 27
Dolcoath Mine 28, 33
Flottman rock drills at 55
man cages at Harriet Shaft 51
new air compressor for 43
pumping engine refitted 52
removal of pumps from Williams' Shaft 114
removal of steam winder from Harriet Shaft 114
repairs to man-engine 50
Williams' Shaft 55
Williams' Shaft winder installed 54
Dolcoath Old Sump Shaft
new headframe 52
Dolcoath Road sports grounds
opening of 177
Dolcoath stamps
engine for 22
Dorothea Slate Quarry
new pumping engine for 67

333

paraffin motor wagon 271
RB440 compressor 279
sliding vane portable compressors 279
Sports Club 280
BroomWade. See Broom & Wade
BroomWade Compressor Hirers 274
BroomWade Power Tools 280
Brown. J. 283
Brown, John 112
Burleigh, Charles 24
Burleigh rock drill 24
  at Rocks and Goonbarrow Mine 30
Burleigh Rock Drill Company 31
Butler, K. 187, 188
Camborne Cricket Club 182
Camborne Public Free Library
  use of room for Holman pensioners 250
Campion, L. 181
Capital and Counties Bank 213
Carbis, W. H. 22
Carn Brea and Tincroft Mines
  new compressor for 66
Carn Brea Mine 30
  Holman compressor for 48
Carn Brea Railway 193
Carn Entral 49
Carn Marth Quarry 199
Carter, J. 188
Carter, R. 188
Carter, W. 188
Castle-an-Dinas Mine
  sand table for 106
Centenary Row East
  demolition of 252
Centralisation scheme 252
Chadwick Patent Bearing Company
  acquired by Broom & Wade 274
Champion rock drill 30, 197
Charles, Miss S. 181
Charlestown Foundry 78, 192
Charter Towers Gold Mine
  Ingersoll rock drills at 47
Chough gun 140
Clay, Dick 274
Clijah Croft 183
Climax Dust Allayer 209

Climax Light Drifter 222
Climax Lightweight Jackhammer 226
Climax rock drill 29, 197
Climax Rock Drill and Engineering Company 35
Climax Rock Drill and Engineering Works Ltd 192, 210
  first meeting of 210
Climax Rock Drill & Engineering Company
  acquisition of shares by Holman Brothers 228
  becomes wholly-owned subsidiary of Holman Brothers 233
  majority of shares held by Holman Brothers 232
  merger with Holman Brothers 235
  first AGM 214
Climax Sleeve Valve Jackhammer 226
Climax Sports and Social Club 179
Clymo, G. 177
Cock, Ben 44, 103, 115, 123, 130
  concrete breaker 124
  death of 148
Cock, J. M. 181
Cock, William 44
Collins, John 188
Collins, Phil 188
Commercial Hotel, Camborne 75
CompAir
  acquisition of title 280
  centralisation scheme for Camborne 310
  demolition of Wesley Street works 309
  end of rock drill production in Camborne 305
  industrial action at Camborne 306
  origin of name 302
  redundancies in Camborne 305
  redundencies at Camborne 307
  sold to Siebe group by ICGas 305
CompAir Auto Power 312
CompAir Construction and Mining 302, 304
CompAir Demag UK 312
CompAir Group
  divisions of 302
CompAir-Holman
  acquired by ICGas 304

Index

Adjustable rock drill cradle
  wins silver medal 210
Admin building
  topping out ceremony 253
Agar, Frank H. 210, 216
Allen, Chris 186
Anglo-American Exposition, London, 1914 92
Angove, B. 181
Angove, J. P. 182
Anti-phthisis device
  in Imperial hammer drill 207
Apprentice necktie, introduction of 161
Armstrong Whitworth 125
Arthur, J. H. 125
Aswan Dam 151

Bain, David W. 50
Ball, E. Bruce 283
Bank of Northern Commerce Ltd 215
Banyard, L. G.
  joins board of Reavell & Co 297
Barrow rock drill 28, 35, 38, 197
Bartles Foundry 158, 193
Bartles, Tregonning & Duncan Foundry 193
Basset Mines 56
  accident at 57
  Marriot's Shaft, engine for 56
  repairs to Lyle's engine 70
Bates, Richard 214
Bates, Robert 210, 213, 218
Batten, George H. 50, 55
Bawden, Jackie 112
B Company 9th (Cornwall) Battalion
  Holman Home Guard Unit 144
Beaumont, Colonel 197
Beaumont, Major 30
Beaumont rock drill 35
  at Carn Brea Mine 30
  fatal accident with 30
Beer Alston Mine 14
Bendixson, Harold 214, 215, 217, 218, 222
  appointed director of Climax 215
Bennetts, Captain William 60

BEN Patents Ltd
  acquired by Broom & Wade 275
Berringer, J. J. 64, 129
Berryman, K. T. 177, 181
Bewick, Thomas
  first foreign order for Climax drill 200
Bickle, Jebus 103
Biddick, W 188
Blackwell, Charlie 120
Blackwell, Harold 28
Blackwell, Harry 120
Blaythorne House 178
  official opening of club house 180
Blewett, Charles 86
Blight, George 197
Blight, Joe 111
Boer War 75
Boilers works
  moves to Roskear 47
Borer No. 45
  completes Mont Cernis Tunnel 26
Bowling Club 179
Brakpan Mine
  Holman sinking record at 82
Bray, Albert 102
Bray, David 321
Bren gun parts
  manufacture of 145
British Air Brake Company 294
British Bank of Northern Commerce 212
British Compressed Air Society 127
British Empire Exhibition, 1924 117
British Pneumatic Tool Company 125
Broad Street House. *See also* London Office
Broom, Dick
  joins Broom & Wade 274
Broom, Harry Skeet 269
  death of 280
Broom & Wade
  history of 269
  manufacture of Churchill tanks 277
  negotiations to merge with Holman Brothers 280

The Holman Film Archive

By the end of 1989 the remaining landmarks of the Holman empire were being demolished. Little of the 22 acre site is now left to identify where 2000 employees worked to invent, develop and manufacture some of the world's most innovative mining machinery.

From the 1920s through until the 1960s Holman's own film unit produced many instructional and informative films looking at the activities of the people and departments within the Camborne factory.

While the factory was being partly cleared in 2003, 160 reels of 16mm film were discovered. In addition to factory processes these reels bring to life the diverse workforce; they highlight drill testing in the mines, safety training, and even capture the little known Holman Projector.

These films have been digitised and the Trevithick Society will be making this material available as a set of three DVDs later this summer.

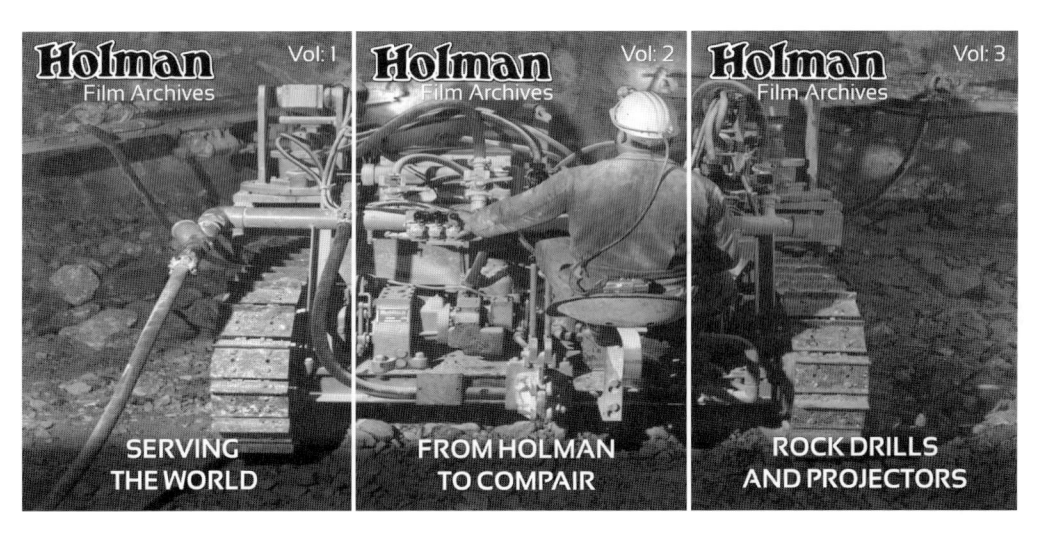

If you would like more information then log on to the Trevithick Society's website at www.trevithicksociety.com where you will find more detail nearer to the launch date. You can also contact us via email at:

pr@trevithicksociety.com

Bibliography

Books

Barton, D. B., *The Cornish Beam Engine*, Bradford Barton, Truro, 1965

Carter, Clive, *Cornish Engineering 1801-2001: Holman Two Centuries of Industrial Excellence in Camborne,* Compair, 2001. Reprinted by the Trevithick Society, 2004. The bicentenary volume.

Hichens, Antony, *Gunboat Command: The Biography of Lieutenant Commander Robert Hichens, DSO*, DSC**, RNVR*, Pen and Sword Books, 2007

Hichens, Robert, *We Fought Them in Gunboats*, Michael Joseph, 1944

Hollowood, Bernard, *Cornish Engineers*, Holman Brothers, 1951. Privately printed to mark 150 years.

Anon, *Fifty Years, some notes of the progress of Reavell & Company from 1898 to 1948.* Reavell and Co., Raneleigh Works, Ipswich.

Carter, Sandra, *A Recorded History of CompAir Broomwade 1898-1998,* CompAir Broomwade

Pamphlets

Climax Illustrated, c1945 The Climax company's war work recorded. Reprinted by the Trevithick Society, 2006.

Our Work during the Great War, Holman Brothers, 1919 Reprinted by the Trevithick Society, 2005.

Souvenir of the Visit of HRH The Prince of Wales, Holman Brothers, 1926 Reprinted by the Trevithick Society, 2005

Tin Mining: Holman Crushing and Dressing Plant, Reprint of a Holman Catalogue of c1914 by the Trevithick Society, 2006

A Souvenir of the British Empire Exhibition, Wembley, 1924, brochure produced by Holman Brothers illustrating the range of the company's activities.

Souvenir of the Empire Mining and Metallurgical Congress, South Africa, 1930, brochure produced by Holman Brothers illustrating the range of the company's activities.

Articles

Carter, Clive, *The Boring Machine: Introduction of Compressed Air Rock Drills into the Camborne Mines,* Journal of the Trevithick Society, No. 20, 1993

shower of rain but that did nothing to deter the crowds. It seemed as though nothing could really go wrong that day.

The replica Puffing Devil has far outlived its predecessor, which only made two trips, and has been shown at numerous events both in and out of Cornwall. The engine appeared at York for the bicentenary celebrations of the railway locomotive where train enthusiasts would not believe that the road vehicle appeared before rail. The Puffing Devil appears at every Camborne Trevithick Day, the last Saturday in April, where the town centre is closed for the country's largest in-town steam parade. The road locomotive takes pride of place at the head of the parade, as senior member of the steam fraternity even though it is not the oldest vehicle on display; it just represents it.

FINIS

steam and smoke issuing from various orifices made it look much faster than it could possibly go.

The long straight road was clear and, although there was little pressure in the boiler (about 25 psi) it was sufficient to travel at a merry trot with the crowds skipping closely behind. This was an historic moment and it was savoured by all that were on board or saw the engine on its journey that day.

A reverential stop was made at the statue of Richard Trevithick outside the library. Another great crowd was assembled and an enthusiastic cheer went up again. Following a gentle turn into Basset Street the little engine was dwarfed by some of its direct descendants in the form of the traction engines which were lined up in preparation for the parade. Basset Street echoed to a tremendous welcome of shrill whistles. The atmosphere was exhilarating.

During all the long months of construction no real thought had been given to the effect the locomotive would have when other people saw it for the first time. In one stroke the little engine had been given life and a family of thousands. It was warm, breathing, chuffing and dribbling. Thousands of people immediately took it to their hearts.

There was little room to manoeuvre in the road outside the Trevithick Surgery car park so the skill of the crew and the agility of the replica were put to the test before the eyes of the public. A deft bit of driving in the confined space was completed to another roar of approval and the locomotive was neatly parked. The pin was removed from the drive and the engine was allowed to idle, chuffing contentedly.

The Society completely surrounded the engine with a screen of John Sawle's sheep hurdles to keep curious fingers off the hot boiler and the moving parts. Everyone wanted to see what Richard Trevithick had produced against the advice of James Watt 200 years ago. All those involved in the project were quizzed endlessly about the intricacies of the engine and how it had been made. Some people just stood silently and marvelled. Cameras clicked incessantly.

The late Frank Trevithick Okuno, a Japanese direct descendant of the Cornish inventor, had made a long railway journey in poor health from London. He was ecstatic that the project had come to a successful conclusion. There was a short

instructions how to act if more marshals were required. Two were designated as handlers of the wheel chocks in case a need arose to use them.

It was quite a distance from the Number 5 Building to the factory gates. The engine was now leaving what had been its home for the last year and making its own way into Camborne. During the previous week period posters had been distributed around Camborne inviting people to accompany the engine on its journey from the factory to the town. Consequently the locomotive was greeted by a great crowd in Foundry Road. With another whistle to acknowledge their waves, cheers and camera clicks the little engine puffed around the Tesco roundabout and was soon on its way along Centenary Street.

The Puffing Devil at Bridgend, 19 July 2008. This event was to celebrate the bicentenary of Trevithick's Catch-Me-Who-Can engine, the world's first passenger-carrying railway locomotive. In the background is the replica of Trevithick's London Road Locomotive. Also present were the replics of the Penydarren locomotive and Catch-Me-Who-Can, the latest Trevithick replica.

From the design stage the locomotive had been intended to have a cruising speed of 3½ mph in order to make it suitable to take part in the Trevithick Day parade and not run away from the dancers. It could clearly travel a little faster but its unusual appearance with a plunging piston, flailing connecting rods and with

pressure was reluctant to rise. Maybe the cold wind that morning was undermining the efforts of the firemen. Time was running out so several CompAir employees came to the rescue with an air compressor (not a difficult item to find in CompAir)! and an air line provided a forced draught.

Soon a working pressure was in sight and the crew slipped away to change into their period costume, an essential part of the re-enactment. Once they were all looking the part a driving pin was inserted into a crank and the engine eased gently forward. With the wheel chocks away and the brakes off, John opened the steam valve and the little engine and its crew were away.

The Puffing Devil passing the commemorative plaque in Fore Street. Kingsley Rickard waves while Arthur Young looks on.

The crew of ten was divided into those being carried and those who were marshals. The loco could be operated by two people, the driver John Sawle, the Project Engineer with 38 years experience of steam locomotives, who was responsible for the engine operation and braking and the steersman, Arthur Young, a skilled voluntary engineer who had built the vehicle and whose sole job was now its direction. All the others were along for the historic ride but had also received

Young set about the tender ministrations required by a steam engine of such an ancient design. Bolts were checked everywhere and the engine was lubricated. The last drops of a thick, dull green oil were used. It had been presented to the West of England Steam Engine Society at South Crofty Mine when Robinson's 80-inch steam engine was laid to rest many years previously.

A small group of CompAir employees chatted merrily and added to the excitement and enthusiasm as Jon Eastman stoked the fire and the final specks of dust were carefully removed from the recently painted black boiler.

The media, always looking for a good story, brought out their cameras, recorders and note books. Much was made of the thousands of hours of skilled work involved in the building of this engine. A group of talented steam engine owners who had arrived in Camborne for the festivities admired the authentic workmanship before them and congratulated everyone they could find. John Sawle and his team were feeling justly pleased with themselves.

"Going up Camborne Hill coming down".

Then the first sign that the day was not going to be without its problems became apparent. Although the fire had been burning fiercely for some time the boiler

realise Richard Trevithick's dreams of his 'road locomotive' or as it should be known, the world's first motor car.

The weekend commencing the 28th April 2001 was to be like nothing else in the history of the Trevithick Society. It started at about seven in the morning when Vice Chairman Kingsley Rickard and his small team erected their incredibly complicated but award winning bright yellow Society tent in the car park at the Trevithick Surgery.

Soon afterwards the engineering team assembled at the Holman factory of CompAir UK to prepare the star of the parade for its important outing. The engine was manhandled into the yard and preparations made to raise steam. There was clearly plenty of time before the planned public appearance at 11 o'clock outside the factory gates.

As the fire was laid and the first smoke appeared, David Bray and Arthur

The Puffing Devil leaving CompAir for the Camborne Hill run..

The Puffing Devil in its bay at CompAir. Note the number plate - the world's first car still needs one!

commodity, known in the American Space Programme as 'The Right Stuff'. The subsequent building of the replica took place in an atmosphere of curiosity and pride, though considerable skill and ingenuity was demanded to reconcile the engineering practices of 1801 to those of a modern plant. The Holman engineers working to the exacting standards of high pressure air compressors were horrified that many locomotive components required 'gaps' rather than tolerances. The greatest advantages of CompAir-Holman hospitality were the ready access to machine tools, the skills and enthusiasm of all those involved and a splendid working environment. Over the coming months their unfailing encouragement and generosity were crucial and the Trevithick Society and the people of Camborne will always owe CompAir-Holman a great debt of gratitude.

In March 2000 the locomotive's boiler arrived from the Deepdale Engineering Co. of Dudley, ready to be mounted on a stout oak chassis which had been precision planed by David Ball Construction of Redruth. Other components like nuts, bolts, wheels and furnace fittings, were added in the coming months, all individually and generously made by many other firms. Such was the progress that the locomotive was displayed in public for the first time on Trevithick Day in April 2000 and, being partly under steam, created an enthusiastic response from townspeople and visitors alike. A year later the 'Roaring, Puffing, Devil' was undergoing road trials in the yard at CompAir-Holman, ready for its maiden run up Camborne Hill in April 2001, two centuries after Nicholas Holman helped

The original Camborne Hill journey of the 'Puffing Devil'.

CHAPTER EIGHTEEN

GOING UP CAMBORNE HILL COMING DOWN

One of the more historic chapters in the history of the Holman engineering empire must be that of building the working replica of Richard Trevithick's Road Locomotive, the 'Roaring Puffing Devil' which went up Camborne Hill on the stormy Christmas Eve of 1801. It was the culmination of four years of hard work by the Trevithick Society, which was the successor of the old Cornish Engines Preservation Society. The Trevithick Society had already brought back under steam the historic Levant winder whose rescue, sixty years before, had inspired the first bid to preserve Cornwall's industrial heritage. To this was added Michell's whim, next to the old A30 at Pool, Taylor's pumping engine behind what is now Morrison's supermarket, and Robinson's pumping engine at South Crofty Mine. Other property included Trevithick's House at Penponds, near Camborne, and the engine house at Wheal Betsy, to the north of Tavistock. All were later given over to the care of the National Trust.

From drawing board to construction had been a formidable task as enthusiasm and dedication had to be matched by an equally vigorous campaign to raise the necessary funds. A large number of Cornish companies were recruited to the project and in 1999 an approach was made to CompAir-Holman. The ensuing discussion could have been held with any of the old 'Holman Boys'. In reply to comment that it was a pity the company no longer had apprentices, Geoff White who later became a Vice President of CompAir-Holman, said, 'We still have plenty of apprentices, except that they are retired and wandering around Camborne but they are still our apprentices'.

Two centuries might have passed but Holmans still retained that glorious

During the last few years, attention turned to the long-abandoned No. 3 Rock Drill Works on the corner of Trevu Road and Trevenson Street, opposite Trevithick's statue. Various plans had been put forward for remodelling the site but it was eventually sold to the property developer Coastline for housing. In 2009 demolition of the corner plot took place as well as the rear of the office building in Trevu Road, of which only the fascia was to be retained. A new roof was put on the old showroom building next to the railway line and repairs carried out to damage which took place in 2008. Unfortunately the 'northern lights' along the railway line were demolished and the distinctive saw-tooth outline, part of the World Heritage Site, has been lost. At the time of writing (Spring 2012) a new roof has been put on the old showroom building in Trevu Road; construction work on the corner plot next to the old Public Rooms had been affected by the failure of a company carrying out the work but this is now complete. Unfortunately the Public Rooms will be demolished because it is in such a dangerous condition; there are plans however to retain the fascia.

The modern CompAir is an international organisation, sadly outside Cornwall. The company now includes a number of well-known global brands:

Rollover	Holman	LeRoi	BroomWade	Cyclon
Reavell	Hydrovane	Kellogg	Luchard	Mahle
Mako	Demag			

Gardner Denver has three facilities in the UK, of which Redditch is the principal; other offices are in Spain, France, Germany, Switzerland, Italy, Austria, Poland, Serbia, Canada, the U.S.A, Brazil, South Africa, Dubai, Singapore, Australia, Hong Kong, South Korea and China.

with manufacturing being transferred to the parent company in Kent.

In preparation for this work, CPR Regeneration quit the site in November and demolition of the PCA and stores buildings commenced on 8th December and was completed by the 23rd. Demolition of the rest of the site, with the exception of the admin block, which was taken over by Royal Cornwall Hospitals Trust, took place early in 2006. The building was demolished at the beginning of 2010 and new homes have appeared on the site.

Quaiffe Engineering departing, 17 January 2006.

In 2008 CompAir-Holman was sold by Alchemy to Gardner Denver, Inc., a leading global manufacturer of highly engineered compressors, blowers, pumps and other fluid transfer equipment, for £200.6 million. As well as the Gardner Denver and CompAir brands the company also includes Belliss & Morcom, Champion, Bottarini, Tamrotor and Air Relief. Manufacture of CompAir products in Europe is based on an 8½ acre site near Redditch, Worcestershire, with 90,000 square feet of factory area and 20,000 square feet of offices. CompAir Industrial Products are produced in Simmern, near Frankfurt in Germany. The only sign of the long history of the company in Cornwall is a sales office at the Cardrew Industrial Estate near Redruth.

PT, empty and roofless, 2 July 2006.

announced an expansion of its operations in June 2004, taking on new staff and introducing 24-hour production. A further expansion was announced four months later with a new two-year contract with General Motors for gearboxes. Quaife was at that time producing 1,200 differentials per month for Daimler-Chrysler.

Holman-OEM led a less successful life and ceased trading in January 2005 with the loss of 25 jobs. The company, making sheet metal fabrications, had experienced cash-flow problems which it blamed on 'previous business forecasts' which had not materialised. All of the company's equipment was sold at auction by English Partnerships in October 2005 while Quaife bought the equipment it previously been leasing. In December 2005 a shortlist of six developers for the site was announced by English Partnerships. Bellway Homes, Crest Nicholson, Midas Homes, Places for People, Taylor Woodrow Developments and Wain Homes were invited to prepare in-depth submissions to create 380 homes, light industrial space, live work units, offices and communal open spaces on the site. Quaife had been under notice to quite the site by early 2006 and closed on February 10th, the year. A management buy-out had been rumoured but this never materialised. However two companies, Quaife and OEM Fabrications, took over parts of the site towards the end of the year; OEM Fabrications formed a subsidiary called Holman-OEM Ltd. Although only a small number of former Holman employees were taken on (12 and 20 respectively) hopes were again raised that engineering would continue on a greater scale. The deal was brokered by CPR Regeneration Company, the Regional Development Agency and English Partnerships; these organisations having bought all of the equipment at the site in order to lease it to the two new companies.

In February 2004 CPR Regeneration commenced holding discussions regarding the future of the Dolcoath area, which included the Foundry Road site. Regeneration of the area was to be a 'mixed-use' neighbourhood based on the Foundry Road site, including light industrial units. In April 2004 CPR Regeneration moved to Foundry Road from the Tolvaddon Energy Park as part of a deal to set up the regional headquarters of a new Government service at Tolvaddon. Quaife

The heart being torn from PCA, 19 December 2005.

Smiles in the PCA stores, August 2003.

had been sold off and workers lost, however it was announced that it was severe global competition and continued trading losses which were the underlying cause and that production would be transferred to Germany. CompAir-Holman's Chief Executive, Nick Sanders, stated

> We have looked at a number of ways to reduce our continued costs at Camborne, including reducing headcount, implementing purchasing cost reductions and introducing initiatives to improve sales revenue. Unfortunately these measures have not been sufficient to counter the effects of poor trading conditions and strong competition and our trading losses continue. Regrettably, due to these adverse conditions, we need to further consolidate our operational facilities and therefore propose to close the site.

Ironically, Sanders had been brought in by Invensys in January 2002 as a troubleshooter 'charged with turning the company around prior to selling it off'.

Eighty five staff were laid off immediately with the remainder to go by the end of

By the following year, CompAir-Holman was being referred to as a 'modernised multi-million pound manufacturing plant' and had installed the only computer-controlled laser cutter for metal plate in the UK. The company was awarded the coveted Ministry of Defence 'AQUAP' designation and continued to expand, though there were still occasional setbacks like the effect that trouble in the Middle East had on the construction markets. New products were added and others phased out while new markets were added, such as Holman compressors providing the power to clean down North Sea oil rigs. Robotic and precision welding enabled the company to produce components in house that had previously been subcontracted and even the headquarters block was refurbished to provide an impressive new frontage.

Atmospheric view of the new admin block, the former stores building.

During the 1990s the company continued the tradition of heavy engineering in Camborne begun by Nicholas Holman two hundred years before, but now it was a different kind of engineering from what was familiar even twenty years before. Men and machines have been replaced by fewer men and far more advanced machines and the days of the crowded, noisy and bustling machine shop had gone for good. Their passing may be lamented by those who can afford the luxury of nostalgia but progress, regardless of casualties, is unstoppable and it is worth

Portable Compressor Assembly.

This dramatic demolition and construction was part of a scheme to centralise the manufacturing facilities even further. Essentially this plan involved the closure of the No. 3 Works (near Camborne Station), the closure of the Drop-Forge (formerly Boiler) Works (north of Roskear), new buildings erected near PT, the sale of the No. 1 Works for redevelopment and the sale of the admin building to Kerrier District Council, the Board Room brought back to the main works after an absence of thirty-five years. This would confine the company to an area of land bordered by Dolcoath Avenue, the new part of Foundry Road, the old part of Foundry Road and the railway line. The new buildings comprised a new 20,000 square-foot R & D facility on a previously unused part of the site just east of the old R & D and a 60,000 square-foot stores/PCA building adjoining the east side of the former Maxam/Plastics building. The Stores and Assembly building, now fronting the re-routed Foundry Road, was converted to an administration building.

as it had been known for a century would be swept away and its demise would help finance the new development by taking advantage of the then continuing boom in supermarket building.

Many welcomed the demolition of the 'drab Victorian buildings' at the entrance to the town and their replacement by a glittering new supermarket, which was Tesco moving from just the other end of Camborne, where the original Camborne School of Mines had already been flattened for the supermarket's benefit. By the winter of 1989 the whole of the frontage of Holmans, including the ornate granite office block and the old foundry, was steadily being pulled down, with apparently no thought being given to the preservation of the historic facade. The last traces of the old railway line to Boiler Works vanished, while the rerouting of the road up to Foundry Road obliterated the old main entrance which had seen the passing of generations of Holmans men.

Construction of the new R&D building; PCA is on the left and PT and the old Central Steel Stores behind.

The stripped interior of No.1 Heavy Machine Shop.

The new redundancies reduced the workforce to 330 with a further 120 employed outside Cornwall. Production schedules and working practices were revised and a vigorous new approach to marketing and development wiped away the last vestiges of the cosy and paternal days of the old Holman Empire. The then Managing Director, Peter Fitch, recognised the value of the name 'Holman' and renamed the company 'CompAir-Holman'. There was an increasing emphasis on the civil engineering market rather than mining, though there was still good business to be had in Sierra Leone, Nigeria and Ghana. The 'Zitair' range of air compressors was further developed and improved which soon resulted in substantial orders from plant hire companies and also from Cornwall County Council.

The works still sprawled over 22 acres and embraced buildings that would have been familiar to old John Holman, standing alongside shops like PT dating from the war and others erected in the 1950s and 60s. Many housed plant and machinery that was either redundant or obsolescent and a radical solution was needed to accommodate a proposed new R & D department and assembly hall, equipped with the latest laser and computerised technology. Essentially the Holman works

By associating yourself with the action you are deliberately causing damage to the company and putting the company under duress.

Anyone not signing this statement or signing it and not working normally will not be paid.

The company is not willing to tolerate such a breach of contract.

Part of the letter from Roy Price, MD

Five hundred workers refused to sign and voted to continue working without pay. At this point Sid Godolphin's job as full-time convenor was scrapped by the management; offered alternative employment on the shop floor, he decided to retire early. The industrial action only lasted a few weeks, the union accepting that the company could not afford 10%; in return the company promised to do all it could to safeguard jobs. This agreement was accepted by 80% of 580 shop floor workers. As if to emphasis the company's position, it made a loss for the previous year, although part of this was caused by payments totalling £750,000 to 100 workers made redundant; 20 jobs were also lost through 'natural wastage'.

Aerial view showing the extent of the works before 1989.

Clive Carter drawing of No.1 Works

CHAPTER SIXTEEN

INTERNATIONAL COMPRESSED AIR AND WORLD DOMINATION

In 1968 the merger of Holman Brothers with BroomWade of High Wycombe and Reavell & Co. of Ipswich to form the International Compressed Air Corporation seemed the path to continuing success, much in the way that Holmans had taken over Climax back in 1951. The merger decision was made in May and was effective on September 30th. The company was instantly Britain's largest manufacturer of compressed air equipment. Another era had ended that same year when Percy Holman ('Mr Percy' or 'Pip') retired as chairman of the Holman Group after serving fifty-three years with the company. If the Holman boys had been a rugby team then certainly Treve would have been the captain and Percy the prop-forward. Sadly Mr Percy saw little of his well earned retirement, as he died suddenly at the age of seventy-three at a county rugby semi-final at Redruth, with Cornwall already beating East Midlands, on 6 February 1969. He had been a life-long Rugby man, having played both for Camborne and Cornwall and been President of the Rugby Union, though he was an equally enthusiastic yachtsman sailing from the Helford River. His business and civic career had been exemplary and he had received the OBE in 1960. Yet Mr Percy seemed to have had more robust adventures than the other Holman boys, such as when, in 1933, seated in the back of a Gypsy Moth, he flew through a tremendous monsoon while his pilot tried to find a course from Karachi to Bombay.

About this time another member of the Holman family appeared on the scene. Richard Carthew Holman, nephew of Treve, was invited to join the company and was put in charge of Research and Development and given a seat on the board. Following the merger with BroomWade he was appointed special projects director. In that capacity, he was involved in the development of such new technology as the Dryductor, a drilling system designed to suppress dust, and in the late 1950s he had been involved with Jim Hodge in designing the then revolutionary Rotair compressor system. In 1945 he had been awarded the Military Cross while fighting in Europe with the Royal Armoured Corps.

In about 1972 the name 'CompAir' was introduced to cover the new organisation; the name was taken from the magazine of BroomWade's apprentices. Efforts were constantly being made to broaden the export market. One problem had been Turkey, which, because of local laws, would not allow the importation of completed equipment. The way around this was to send 'kits' to the country to make up complete compressors and rock drills. Rock drills were also used to hook customers, as a rock drill 'would eat itself in six months', following which it would require regular replacement parts.

Dick Holman.

In the early 1970s the companies within the CompAir Group were re-organised under product divisions:

- CompAir Industrial, manufacturing BroomWade and Reavell industrial compressors, tools and other equipment
- CompAir Construction and Mining, manufacturing BroomWade and Holman portable compressors and all contractors and mining tools
- CompAir Maxam, manufacturing controls and pneumatic circuitry
- CompAir International, which unified the overseas operations

By 1975 the CompAir Croup was the largest manufacturer of air compressors and pneumatic equipment in the UK. World sales for 1975 totalled £86m, with pre-tax profit of £7.3m. The group employed more than 8,000 people, ran seven manufacturing locations in the UK and had subsidiary companies active in 16 markets overseas.

James F. Holman, had already taken over as chairman of the Holman Group after Mr Percy's retirement, but he died in December 1974 at the comparatively young age of fifty-eight. He had spent his youth mining in Canada and after service with the RAF had taken over as the company's London director in 1950. He would undoubtedly have taken the modernisation of Holmans even farther.

Even as the Holman family slowly relinquished their command of the company formed by their great grandfather, these were still the days of the 'Foundry families' sons who had followed their fathers and grandfathers into the foundry. Some had extraordinary records of service. The Trebilcocks accumulated over 300 years of combined service but few could rival the Godolphins, who stemmed from Francis Godolphin, an ex-Harvey pattern maker who joined Holmans back in 1896 and whose sons and grandsons served the Holmans for over a century.

The Godolphin family, 24th October 1972
BACK ROW: W. Terence Godolphin, Paul "Sammy" Godolphin, R. Sidney Godolphin, Francis "Charlie" Godolphin, R. Owen Godolphin, Ronald Godolphin, Robin J. Godolphin

FRONT ROW: K. Malcolm Godolphin, Leonard Godolphin, Mrs "Carrie" Bullock, Mrs Ann Lawson, W. Kenneth Godolphin, R. "Terry" Godolphin.

Yet the optimism, even enthusiasm, of the seventies soon evaporated and merely became a prelude to the near destruction of heavy industry in Britain. Much of the damage had already been done and was often self-inflicted. A hankering for the old days of easy trading with a captive empire, a supine refusal to define the proper role of trade unions, the delusion of continuing superiority in foreign markets and the failure to recognise the potential of the European market, all contributed to the greatest re-shaping of British industry since the Industrial Revolution. CompAir Construction and Mining, although the world's third largest manufacturer of compressed air equipment and with a far better record than most companies, could not escape unscathed by the unforgiving economics of the 1980s.

In 1979 a strike took place which had unforeseen consequences. June that year marked the retirement of Dick Holman from CompAir, marking the end of the family's 178-year connection with the firm. At the beginning of May a strike was called because of the terms offered to those workers taking voluntary redundancies. Three unions were eventually involved and the strike finally ended in July.

As Dick left the Works he had to pass through the picket. The strike committee then explained apologetically that because of the strike they had not been unable to arrange a collection, but they gave him a card signed by all 380 members of staff on strike and was given a rousing three cheers. After the strike ended a collection and a touching presentation were made.

In 1980 CompAir-Holman was taken over by ICGas at the very start of this unhappy period which had already seen the demise of Alfred Herbert Ltd, the supreme British machine tool maker, where Treve Holman had served his apprenticeship so long ago. The company's name was changed to 'CompAir', meaning that for the first time for over a century the name Holman had not been used. Other familiar names would soon be extinguished as the deepening recession struck in housing and civil engineering, seriously affecting the makers of construction plant and machinery. CompAir's catalogue still offered a very wide range of products and was a publication that would have delighted the old Holman Boys who, significantly, would have easily recognised the new generation of Silver Threes, P. W.s and Road Rippers, despite the innovation of noise mufflers.

In December 1980 a new works manager, Mr. Andrew, was appointed. His instructions from ICGas were to put CompAir's finances in order. One of the possibilities he considered was the total closure of the factory but instead opted for a complete reorganisation of the CompAir's business. The first step was to reduce the number of managers (to three), then each of the managers was required to "pick a team" for his department so that staff were retained to keep the company viable. Unfortunately this led to redundancies and, later, a court case resulting from an industrial tribunal.

1985 was to be a year of dramatic and radical change for what remained of the Holman Empire. After years of heavy losses, CompAir was at last marginally profitable but by February there were alarming rumours that the Camborne works might close and the whole operation might be transferred to Marlow; the compressor plant at Desborough having already shut down. What ensued, to use a once popular phrase, was an 'agonising reappraisal' of the company's entire future.

By the summer, the inconceivable was being proposed that rock drill production should cease after 105 years. CompAir equipment was still being made to withstand 'enormous punishment' and be 'exceptionally reliable', a legacy of the days when Holmans proudly boasted that everything they made included 'one third for efficiency'. It took an hour and half to forge a rock drill cylinder while the Japanese made the same component from a steel pipe, producing a machine which was just as effective and also cheap enough to be discarded at the end of a contract. The days of the Silver Bullet and the Silver Three appeared to have ended and underlined the unpalatable truth that survival relied upon rationalisation in production, marketing and development.

The painful process began in September 1985 when ICGas, which wished to concentrate on their oil and gas business, sold CompAir to the Siebe Group, an acquisitive and hard edged conglomerate with a large engineering sector. Siebe was formed in the 1820s through the inventiveness and business skills of Augustus Siebe (1788–1872), a German-born British engineer chiefly known for his contributions to diving equipment, having designed the first ever diving suit, prior to 1830. Rock drill production virtually ceased and within weeks came 500 redundancies, at a time when most large employers, including the last tin mines, were already in deep trouble. The effect was similar to that of the closure of Dolcoath Mine back in 1921 but now there was little or nothing to cushion the devastating loss of jobs and the effect this would have on the town of Camborne.

December 1986 saw the loss of a former Holman employee, David Penhaligon MP, killed in a car crash on the 22nd. Penhaligon had joined the Liberal Party in 1963; he led the Truro Young Liberals and built up the local party (which had been the weakest in Cornwall) into one of the strongest and was the chair of the Cornish Young Liberals from 1966 to 1968. However he was not selected as Liberal candidate for Truro in the 1966 general election (nor for any other seat) and was also rejected for Falmouth and Camborne in 1968 apparently because his strong Cornish accent was thought unattractive.

In the 1970 general election he fought the Devon constituency of Totnes when the previous candidate, Paul Tyler, transferred to Bodmin. He polled poorly in the context of an election in which the party as a whole suffered. However, Penhaligon had acquired useful experience of fighting election campaigns and picked up additional tips from Wallace Lawler's practices in inner-city Birmingham. Penhaligon was first elected to Parliament in 1974; from 1983 he headed the Liberal by-election unit which planned the campaigns in individual seats. At the Liberal Assembly in September 1984 he was chosen as President-elect of the Liberal Party (the first sitting MP to be elected to the post), and served as Party President from 1985 to 1986.

He was appointed as Chief Spokesman on the economy from 1985, the last post he was to hold. At 6:45 on the morning of 22 December 1986 he was travelling to a constituency engagement visiting the Royal Mail workers on the Christmas post at St Austell Post Office, when a van skidded on an icy road and hit his Rover SD1 car near Probus. Penhaligon died instantly; ironically, investigations showed that Penhaligon had not been wearing a seat belt, the recently introduced safety measure for which he had voted in Parliament.

The Holman work force had been shrinking over the years from about 3,000 in 1962 to 2,275 in 1974 until in 1984 only about 850 people had remained at the Camborne factory. Industrial action at the beginning of 1985 caused ill-feeling between workers and management. At the end of 1984 Sid Godolphin, the union convenor, put in a claim for a 10% pay rise to commence at the beginning of 1985. The management claimed lack of funds, the company only looking to break-even, and refused. Union members then started a work-to-rule and an overtime ban. The management responded with a retraction of the bonus scheme and by sending a letter to all staff members requiring them to sign a pledge to work normally.

In 1900 Arnold Hills decided to expand his business interests by acquiring the engineering firm of John Penn & Sons. In order to raise new capital to finance the takeover, he decided to make the Thames Iron Works a public company. This meant that he was no longer in a position where he would be allowed to pump company money into the football club. As a result of this move, the football club was also reorganized and Thames Iron Works FC became West Ham United FC. West Ham began to verge on the edge of bankruptcy and by the end of the 1903-04 season the club had only enough money to pay the wages of one professional player, Tommy Allison, through the summer; sadly, 107 years, later this situation seems not to have changed greatly.

enough to hold the people who wanted to see their performances.

Postscript

Arnold Hills was a charismatic and caring man. Born on the 12th March, 1857, he was educated at Harrow and played for the school's football and cricket teams. He was a talented footballer and represented Oxford against Cambridge in the varsity match. Hills was also the AAA one-mile champion while at Oxford and played for the university in the FA Cup Final in 1877, though on the losing side. An inside-right, Hills was for many years a member of the Old Harrovians and in 1879 won an international cap playing for England against Scotland, a game that England won 5-4.

In 1880 Hills joined the board of his father's company, Thames Ironworks & Shipbuilding. He initially lived in the East India Dock Road in Canning Town. He became concerned about the living conditions of the local people. Hills commented that "the lack of recreational facilities was one of the worst deprivations in the lives of West Ham residents". He added "the perpetual difficulty of West Ham is its poverty, it is rich only in its population."

Arnold Hills, a poor quality image from the Vegetarian Magazine of 1889.

In 1888 Hills became President of the London Vegetarian Society. He also served as President of a London Vegetarian Rambling Club and founded the magazine, *The Vegetarian* and was also an active member of the Temperance Society. Hills established the *Thames Ironworks Gazette* in 1895. It was an amalgamation of local newspaper, popular history magazine and company newsletter. On 29th June, 1895, Hills announced in his newspaper that he intended to establish a football club. From 1899 the Thames Ironworks Football Club started to employ professional players.

and Lieutenant-Colonel K. Reavell became joint managing directors. Quertier was responsible for the technical side of the work, including design and research, while Reavell continued as sales manager and supervised the running of the works. Quertier died on the 21st April 1958 after short illness following a trip to South Africa.

From single and two stage compressors the company moved on to a wide range of more modern types, such as sliding vane rotary reciprocating compressors up to 6,000 psi and boosters to even higher pressures. Centrifugal compressors are manufactured for industrial, marine and civil applications. Reavell also designed and manufactured high pressure compressors for warship and defence applications. These compressors are supplied to the Royal Navy and to numerous others navies such as those of Australia, New Zealand and Canada. Centrifugal compressors are also supplied to liquid gas tankers. These collect boiled-off natural gas then compress it and deliver the then liquid gas to dual burners in the main propulsion boilers. The gas entering the compressor is below -161°C and leaves it at near ambient temperature, a remarkable feat. In addition the company designs and manufactures a wide range of compressor mountings. For all of the design and testing Reavell invested at an early date in computerised processes, particularly computer aided design.

When the First World War ended in 1918 a victory fete with athletic sports and tea was held on a hired field, and this was so popular that as soon as the new sports field was ready the sports meeting became a regular annual event every year. A Christmas party was given in the canteen for the children of all employees with tea and some form of entertainment, and presents, which were a very popular event.

In the earliest days there was an enthusiastic cricket club and works teams were got together for various games, but it was not until 1920. when the company acquired a piece of land on the outskirts of the town and turned it into a fine sports ground with tennis courts and bowling green and built a pavilion, that the works teams could play on their own ground. The Ranelagh Works Athletic Association was at once started with Gaskell as its first president, and sections were formed for cricket, football, hockey, tennis, and bowls, other sections being added later. The Association also promotes indoor entertainments, using the works canteen for dances, concerts, and dramatic performances, though recently the dramatic section has had to take a larger room in the town, as the canteen was not large

a combination bed-plate. Another interesting piece of work which was undertaken was the design and manufacture of the pneumatic steering-gear for target boats which carried no crew and were navigated by wireless control from a distance.

The end of the war did not make Ranelagh Works any less busy as, though the war work ceased and some contracts were cancelled, a good deal of civil work was to be done, held over from the war. Unfortunately the machinery had been run ragged during the war and orders came in much faster than the works could execute them. As a consequence, delivery times were greatly extended. A large proportion of these orders were for export.

After the war the company concentrated on producing new types of compressor, and there was a great demand for compressors to produce absolutely oil-free compressed air and gases, especially for chemical works. This resulted in the design of vertical, single- and two-stage compressors using carbon piston-rings and gland packing, which require no cylinder lubrication. Another important addition to the many types now built is the axial flow turbo compressor.

Reavell compressor on display at Poldark Mine, Wendron.

Following the death of Hugh Paul in 1947, Sir William Reavell became chairman and managing director, and L. G. Banyard, who had been secretary since Gaskell's death in 1931, and D. A. Palmer, who was managing director of Joseph Rogers & Sons Ltd, joined the board. Palmer received his engineering training at Ranelagh Works and had been an assistant to William Reavell from 1909 to 1914. Sir William died early the following year and H. A. Hartley was elected chairman, but sadly died only two months later. D. A. Palmer was then elected chairman with Quertier as deputy chairman; Quertier

Large numbers of high-pressure hydrogen compressors were built for the RAF for compressing hydrogen into cylinders for inflating barrage balloons, and many oxygen compressors for charging the cylinders of the breathing apparatus for high-altitude flying.

Large numbers of high-pressure air compressors were supplied to both the

Reavell three-stage compressor.

Admiralty and the Air Force for torpedo charging. Two types were supplied, both vertical four-stage machines. Compressed air was required for operating the pneumatic brakes and clutches in tanks, and two new compressors were specially designed for this work and thousands of them supplied. These were two-stage, single-acting, air-cooled machines, arranged for direct driving from the main engine. Two new types of compressor were also designed for the Air Force for servicing trolleys for aircraft.

Another new compressor was produced for charging the starting-bottles on the Diesel engine-driven *Hunt* class destroyers and motor patrol boats. This machine was arranged for direct attachment to a small Lister-Diesel engine. The compressor, having no main bearings of its own, was driven by a crank disc fixed on the engine shaft. This made a much more compact set than the earlier arrangement, in which a complete compressor had been coupled to the engine on

Air Brake Company and the Consolidated Brake Company, to sell the complete brake equipment with compressor or exhauster made by Reavell & Company and the other parts by G. D. Peters. For the compressors the single-acting vertical type was used with two unjacketed cylinders, the machine being specially built for direct attachment to an electric motor so as to make a light, compact unit. For the vacuum brakes the rolling-drum rotary made a very satisfactory exhauster. Large numbers of both types have been sold for use on electric trains, tramcars and trolley buses, and more recently on motor-buses and commercial vehicles.

In the 1938 New Year Honours List, William Reavell was appointed Knight Bachelor (KB) in recognition of his many national services.

The shadow of the Second World War was felt at Ranelagh Works some time before it actually started, as for some years the company had been building torpedo-charging compressors for use on warships, including submarines, and their single-acting vertical machines had been adopted for general service compressors on battleships and cruisers. The augmented naval programme, in view of the possibility of war, meant that large numbers of these machines were required, and standard types of compressors were also needed for the shadow factories which were being put up for producing aero engines and aircraft. Most of these factories were modelled on existing works for which Reavell & Company had supplied the compressors, so that naturally the same make of machine was asked for. The works were, therefore, already very busy when war was declared, and all through its duration were working night and day trying to supply all the various compressors which were asked for.

The works were very fortunate in escaping damage by air raids, though there was a very anxious moment when a damaged German bomber missed the end of the erecting shop by inches in its descent. Underground shelters had been constructed and the RAF used to send a special warning when an air raid was imminent, so that work could continue until the last sate moment, leaving just time for everyone to reach the shelters before the raiders could arrive.

The works provided its own fire brigade and fire watchers, and a company of the Home Guard was formed in the Ipswich Battalion, raised and commanded by Lieutenant-Colonel K. Reavell, this company being manned and officered entirely by employees at Ranelagh Works.

of the British pits for that type of work, so that no great number was sold. A modified form of the pick, however, proved to be very useful for other purposes, such as the demolition of brickwork and concrete of moderate thickness, also when fitted with a spade instead of the usual moil point, for digging in hard clay. Many hundreds of these tools were made for the Army during the Second World War, arranged for use as picks or spades.

There was another market which up to that time had been almost completely in the hands of a company selling American machinery. This was the supply of pressure and vacuum brakes for managing railways, including the compressors or exhausters for operating them. Reavell & Company was prepared to make the compressors and exhausters, but did not want to make the brake gear, which was not work which suited them, so they entered into an arrangement with G. D. Peters & Company Ltd., of Slough, who started two subsidiary companies, the British

1940s high pressure compressor for charging torpedoes.

neglected, and there was a large demand for portable petrol-engine driven compressor sets for breaking up the old surface so that they could be remade. The concrete breakers used for this purpose and the portable compressors for operating them were at that time usually imported from America. Reavell & Company had built their first petrol-engine driven portable compressor in 1905, and they were soon building the compressor sets which were wanted, and also made the concrete breakers for use with them. They continued to build portable air compressors with petrol or diesel engines until the Second World War, when they were obliged to give them up to make room for more urgent work.

In 1928 R. L. Quertier, who came to Ranelagh Works in 1906 and had been in charge of the company's London office since 1911, joined the board of directors, and in the same year Captain Kingsley Reavell, the only son of the managing director, also joined the board. After completing his training in the works, he had served with the Forces in the First World War, returning to Ranelagh Works in 1919. William Reavell becomes President of the IME in this year.

The year 1929 was not a good one for business generally, and the effect of this was to give the company the opportunity of adding about 50 per cent to the area of the shops without seriously disorganizing their work. This new extension included a new erecting shop, 40 ft. wide and much higher than the original shops, to enable much larger machines to be dealt with, new stores built across the end of the machine shop with an upper floor for works offices, and a new tool-room and jig stores. A strip of land between the works and the river was purchased to make room for these extensions, and in 1930 about three acres more were acquired for future extensions on the other side of the railway. This land was low, and being in the bend of the river was subject to flooding, but it served as a convenient dump for ashes and spent sand from the foundry, by which the level of a great part has been raised well above flood level, so that it is now ready for new buildings as soon as they can be put up. A subway has already been constructed under the railway line to give access to it.

About the same period the Reavell-Mossay Pneumatic Tool Company was started with the idea of manufacturing and introducing a Belgian patent pneumatic pick into the British collieries, Reavell & Company manufacturing the picks in the shop which had been used for making shell fuses, and Mossay & Company Ltd. doing the selling. The pick, however, was not very well received, either on account of the prejudice of the miners against anything new, or the unsuitability

compressor except to old customers, who appreciated its efficiency, so it was completely redesigned in a new series of sizes up to 1,200 cfm, known as the "Q" class, and had many new features provided to make it foolproof and suitable for running at higher speeds. It was also decided to make a cheaper type of machine for customers who would not pay the price for the quadruplex machines, and the company therefore started to make the simple single-stage, single-acting, vertical splash lubrication compressor in sizes up to 300 cfm, of which thousands have been, and are still being supplied.

The single-stage, single-acting type was not considered suitable for larger capacities on account of the difficulty of getting rid of the heat of compression. A line of two-stage, double-acting vertical compressors had been designed, and some sizes had been made as early as 1910, including some steam-driven machines of about 1,750 cfm capacity, for continental shipyards. These were now brought up to date, and provided a series of compressors for capacities of 500 cfm and upwards for pressures up to 120 psi. Among the many users of this type of compressor was the Ford Works at Dagenham which elected one size as their standard compressor, and had about fifteen of them in use.

The rolling-drum rotary compressor, which has already been mentioned in connection with the ballast tanks of submarines, was designed in several sizes and provided a very useful series of machines for pressures up to about 20psi and for capacities up to about 3,000 cu. ft. per minute.

Just before the war, the company had arranged to acquire the British Empire rights in a patent by the German Aerzener Company for a special type of turbo compressor. William Reavell had a narrow escape from being interned in Germany for the duration of the war, as he had gone to there to sign the agreement and had only left the day before war was declared. The first two of these turbo compressors had been built in 1916, but it was not until after the war that they began to form an important part of the output of the works. Since that time the turbo side of the business has developed enormously, and new types have been designed for various duties including some for smaller capacities than had previously been considered suitable for turbo compressors. Turbo machines have proved particularly suitable for gasworks' use and it is satisfactory to note that about 95 per cent of all the London gas passes through Reavell turbos.

During the war the streets in London and other cities had been to a great extent

In 1916 an extension was necessary to the offices, and this was built with a canteen above it to provide meals for the workers. The works ran day and night continuously until the end of the war in 1918.

The end of the war brought the cancellation of Government contracts for large amounts of equipment for war purposes, and new orders had to be found to take their place. During the war several new competitors had appeared who were making single-acting vertical compressors, which though not so efficient as the Reavell quadruplex compressor had the advantage that they were much simpler and cheaper to make, which was an important matter at that time, when all manufacturers were trying to get their factories back on to a peace-time basis as quickly and cheaply as possible. This made it difficult to sell the quadruplex

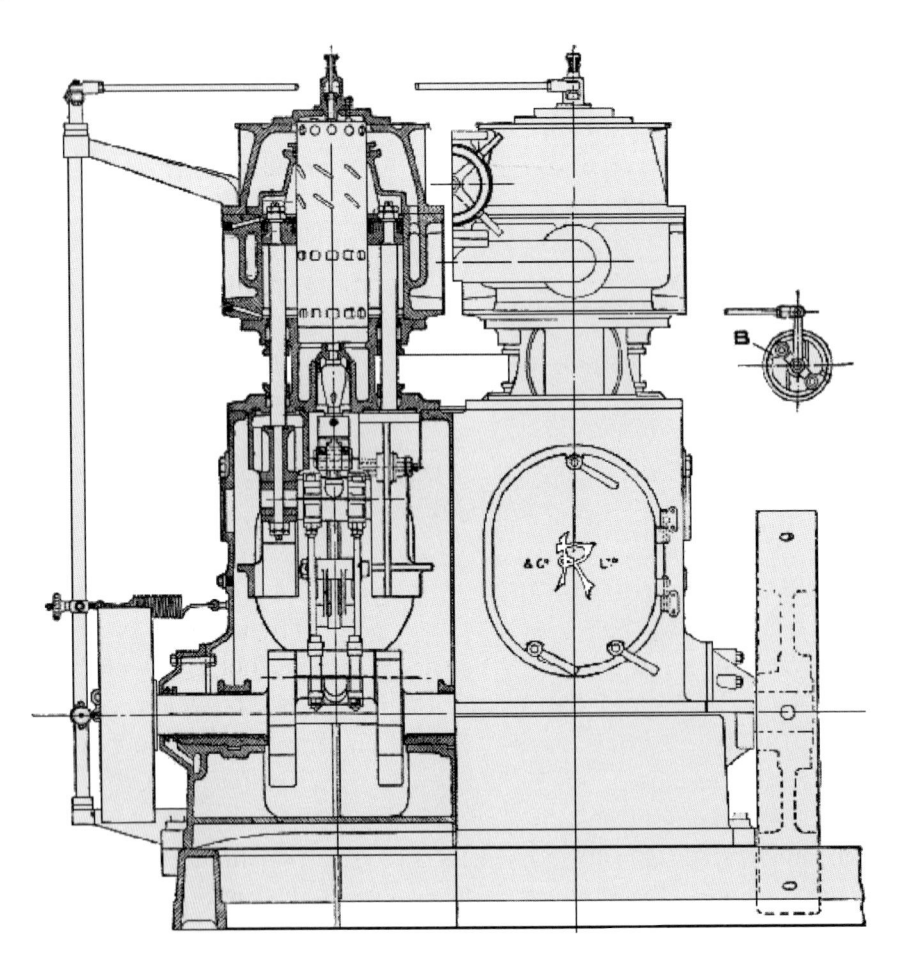

Longitudinal section through a Reavell high-speed compressor.

When the war started in 1914 the works were asked to make 18-pounder shells, but it was very soon decided that they might be much more usefully employed. The company was appointed by the Admiralty to be repairers for the submarines based at Harwich. A shed was put up at the Ipswich Docks, and any submarine which required overhaul or repair used to come up from Harwich so that it could be dealt with. Some of the earlier jobs to be done consisted of replacing the star clutches, and it was gratifying when, as a result of this work, the company was asked to make the star clutches for the new submarines which were being built. Later, several different types of hydroplane gear for submarines were made, incorporating improvements suggested by William Reavell.

A few rotary compressors had been made before the war. These were of the crescent type with the rotor and blades enclosed in a perforated drum, rotating freely on its own bearings with a little clearance inside the casing, as still made. One of these machines was tried in a submarine for discharging the water from the ballast tanks, and as it was found that it could not be heard nearly so far away as the piston-type machine then in use, it was adopted for the duty, and large numbers of them were supplied.

For the War Office three special types of hand-operated compressor were designed for working up to 700psi for charging the recuperators of guns and howitzers, and after satisfactory types had been produced, several thousand were supplied.

In 1915 a new three-story building was put up for producing shell fuses, the work being done by women with only a few men acting as tool setters. There were about three hundred women working in two shifts, so that the work was continuous, and some thousands of fuses were turned out each week.

Towards the end of the war the same shop was used for making "Gray-Wimperis" gyroscopic bomb-sights for aeroplanes, the gyroscope being driven by an air turbine running at 25,000 rpm, developed by Reavell & Co. which was supplied with air by a rotary compressor fixed under the plane and driven by a small windmill.

In addition to this special work, there was also a large demand for compressors of standard types for other firms having war contracts, and many three-stage compressors and some five-stage machines were supplied to the RNAS for compressing hydrogen into bottles for airships and observation balloons.

Various other types of compressor were built during this period, including horizontal and vertical double-acting machines, some of these with steam cylinders incorporated, and in 1908 the "Axial" type was produced. This was a very simple machine with no connecting rod. There were two horizontal cylinders face to face with a double-ended piston having a vertical bore in the middle in which worked a turned cross-head which had a bore at right angles to its axis to receive the crank-pin. The working pressure for this type of compressor was limited, but it was cheaper to make than the quadruplex type, and large numbers were sold.

In 1907 William Paul, from an Ipswich firm of corn merchants and maltsters, joined the board of directors, and was elected chairman when Arnold Hills retired in 1912. He resigned the chairmanship in 1924, but remained a director until his death in 1928. His son, Hugh Paul, joined the board in 1922 and was chairman from the time his father resigned until his death in 1947.

In 1908 oil engines were added to the company's products. These took the form of petrol-paraffin engines, some of which were sent abroad for driving centrifugal pumps for irrigation, but most were used for country-house lighting. These did not prove a great success, as continual visits were needed to show the gardener or odd-job man—who usually had charge of the engine—what he had done wrong, and this more than absorbed the profits made on building the engines.

The quadruple compressor for pressures up to 100 psi, which still formed the bulk of the output of the works, was completely redesigned in 1907 as a single-stage machine in sizes up to 500 cu. ft. per minute for singled-ended machines or twice that capacity when double-ended, and with only minor alterations this remained the standard Reavell compressor until the end of the First World War.

This new form of quadruple compressor proved very satisfactory and with the demand for these machines and the high-pressure compressors for diesel engines it was necessary to extend the works. In 1908, 90 ft. was added to the length or the shops and more office accommodation was also provided. In 1909 a new brick chimney-stack was built to replace the original iron chimney, and in 1910 the drawing office was more than doubled in size. In 1911 a two-bay foundry, with two cupolas, two coke ovens, and a sand blast chamber, was put up to enable the company to make its own iron castings, and in 1912 the length of the main works was again increased.

built for capacities up to 600 cu. ft. of free air per minute, and was fortunate in securing orders for a large number of these machines for the construction of the London tube railways. From that time the air compressor was firmly established as the company's principal product, and it has remained so ever since, though its form has been altered from time to time to suit new requirements.

In 1905 there was an important new development in the business. An experimental air compressor had been built with the idea of developing the quadruples compressor as a three-stage machine for working at a pressure of 3,600 lb. per sq. in. for torpedo charging in the hope that the British Admiralty would adopt it in place of the leather-packed machine then in use, which required frequent renewal of leathers. The machine itself was not a complete success, but it provided just the experience which was required to enable the company to satisfactorily answer Dr. Diesel's appeal in 1905 for a compressor which would run continuously at a pressure of 1,000 psi to supply the injection air for his pressure ignition engine. The machine produced for this duty was a three-stage quadruplex compressor, with two first-stage cylinders horizontal, one second-stage cylinder at the bottom, and one third-stage cylinder at the top. All the four pistons were driven by a single crank-pin which was attached to the end of the diesel engine crankshaft, the compressor casing being attached to the engine-bed. The cylinders with their valves and the intercoolers between the stages of compression were all enclosed in an ample water-jacket. This machine proved to be exactly what was required, and large numbers were supplied to Carel Freres of Ghent, Willans & Robinson of Rugby, and many other builders of diesel engines up to the time when air injection was generally replaced by pump injection and compressors were only required for charging starting-bottles and manoeuvring, for which simpler and cheaper machines were suitable.

Another type of compressor introduced in 1909, of which many were supplied for the same duty, was a novel design of three-stage machine, known as the "V" class, in which the second stage had no valves, the first- and third-stage cylinders were in tandem at the top, and the second stage at the bottom. The idea was that on the upward stroke the volume in the first-stage cylinder was compressed into the smaller volume of the second-stage cylinder and the intercooler. The pressure was retained by the first-stage delivery valve, and on the downward stroke the air was further compressed into the smaller volume of the third-stage cylinder and the intercooler, the air which had passed the third-stage suction valve being discharged through the final delivery valve on the next upward stroke.

still running and giving good service in a paint works near London until they were "blitzed" in the Second World War, shows the quality of the Reavell product.

These first compressors were the quadruplex machines designed by Reavell, which had four cylinders arranged radially in a circular casing. This casing was cast with an annular space through which the cylinders passed, so as to form an ample water-jacket, and a common passage round the cylinder heads to receive the air discharged through the multiple delivery valves. These valves were light steel valves, spring loaded, with gunmetal seats arranged round the head of each cylinder. The pistons were of the trunk type, and the four connecting rods were driven by a single crank-pin. No suction valves were used. The central part of the casing was the suction chamber as well as the crank chamber, and air was admitted to the cylinders through ports cut in the hollow gudgeon of the connecting rod which coincided with similar ports in the top of the piston during the suction stroke and closed as the connecting rod swung over at the end of the stroke. This feature was largely responsible for the very high efficiency of this type of compressor.

Double-acting vertical compressor, circa 1912. The great size of this machine is shown by the ladder on the right.

With the increased use of rock drills, coal cutters in collieries and other pneumatic tools, there was a demand for compressors of greater capacity, and in 1902 the company brought out its tandem two-stage quadruplex compressor, which was

steam engines were built for electric lighting and for driving the forced draught fans which were also made in the works at that time. The company had been put on the Admiralty list and secured an order for thirty-six sets of fans and engines for three battleships which were being built by the Thames Iron Works. A few air compressors and some pneumatic hoists were also made at this time.

Brown left at the end of 1899, and his place as works manager was taken by W. Jones until early in 1901, when he left and Bruce Ball took over the duties of works manager, E. W. Jones being engaged to take his place as chief draughtsman. H. K. Finch became a director in 1900 and was for a short time at the works as assistant to William Reavell. Bruce Ball left in 1903, and H. A. Hartley came as works manager. He joined the board of directors in 1908 and continued as works director until his death in 1948.

When the first Ipswich power station was built in Constantine Road, to provide power for the new electric trams which came into service in 1903, it was Reavell's who took on the job of building six steam generators that were installed in 1902 at a cost of £6,564. The units drove German built dynamos made by the Allgemeine Company of Berlin. Reavell's works was within site of the town's power station and the workers could proudly see the power station, which provided electricity for the town. Again a local firm was used to build the power station, and workshop. Kenney's built the station at a cost of £26,948 for the generating station and £6,704 for the car shed. This impressive looking building is still partly used by Ipswich Buses.

The introduction of forced lubrication made it possible to increase the speed of double-acting engines, so that they were suitable for direct coupling to electric generators, the Scott engine, which was virtually a single-acting engine, could not compete in price, in spite of its high efficiency and low steam consumption. Orders for Scott engines therefore fell off rapidly, and although the company continued to build small double-acting steam engines, including a large number for lighting the L. C. C. Thames Steamboats in 1905, it was production of air compressors which took off and became the stalwart of the company.

A number or compressors had already been made, and these at first were all single-stage machines, of which there were six sizes, varying in capacity from 17 to 160 cu. ft. of free air per minute at 100 psi. The fact that Compressor No. 1, which was a machine with cylinders 6 in. diameter by 6-inch stroke built in 1899, was

received a bonus equal to half the saving he made by reducing the time taken.

The Scott engine differed from other steam engines in being a compound engine in which both stages of expansion took place in the same cylinder, and though double-acting, exerted a constant downward thrust on the crank. The crankcase was enclosed and splash lubrication used, the parts being in constant thrust making this type of lubrication especially suitable as a large area of the bearings could be left uncovered by the retaining caps. The piston was annular in form and had two piston rods, each with its own crosshead, these being connected by the gudgeon-pin. The valve worked in a ported liner which passed through the centre of the piston and was driven by a bell-crank pivoted in the crank chamber and connected at one end by a link to a pin near the centre of the connecting rod and to the valve spindle at the other. The high-pressure steam was admitted to the upper side of the piston at the top of its stroke, the lower side being open to exhaust, and when it reached the bottom of the stroke the exhaust valve was closed and ports opened connecting the spaces on the two sides of the piston for about half the upward stroke. At that point the transfer ports were closed and the steam below the piston expanded doing work while the steam which was left above the piston was compressed to its original pressure in the clearance space provided for the purpose.

This very much reduced initial condensation and was the chief reason for the high efficiency of this type of engine. The Erst engine, which soon had to be supplemented by a two-crank engine of twice the power, was a single-crank engine of only 60 hp and was soon completed. Some orders for engines had been secured so that the company was able to start production in earnest and numbers of engines with one, two, or three cranks were built. At that time electric light was coming into general favour, and municipal power stations were being put up in all cities and towns of any size. The Scott engine was particularly suitable for driving electric generators, as its special construction enabled it to run at higher speeds than most other engines of that period, so that it could be directly coupled to a generator running at a reasonable speed. The company was kept busy for about four years building these engines, and equipped power stations for Ipswich, Barnstaple, Dartford, Horsham. Heckmondwike, and many other cities and towns, as well as supplying engines for a large number of factories and large stores for driving generators for electric light and power.

In addition to the Scott engine a large number of single-acting and double-acting

had been engaged as works manager, and enough modern machine tools of British and American make had been purchased to enable the company to start building its first Scott engine. This was to drive an electric generator to supply the power and light for the works, which were being served temporarily by a second-hand Allen engine which had been bought for the purpose.

Reavell quadruplex compressor.

The working hours adopted were unusual for the UK in those days, as it was then customary for men to work for two hours before breakfast. It was realized that hungry men tire quickly and are not highly efficient, and that they would work much better if they had breakfast before they came to work, so the hours were fixed at 7am till midday and from 1pm till 5.30pm, work finishing at midday on Saturday. This meant that there were 52½ working hours in the week, but the men were paid for 54 hours if no time had been lost during the week. At that time it was usual to pay either a fixed rate per hour, or a fixed price for the job, but at Ranelagh Works the premium bonus system was adopted from the start, and is still in use there, by which a standard time was fixed for the job and the man

were purchased on the understanding that more could be acquired when it was needed for future extensions. As the site was between the Ranelagh Road and the River Orwell, and there was already an Orwell Works in the town, it was decided to call the new factory the Ranelagh Works.

Fortunately, the site taken was clear of pre-existing buildings and the company was able to build the most up-to-date works that it could, incorporating the newest ideas from the United States and the Continent. The result of this was not only that the shops were of the most modern and efficient type when they were put up, all electrically driven with electric lighting and heated by the exhaust steam from the engine which drove the electric generator, but the original buildings with very little alteration were suitable to form part of the present layout, which can be regarded as a natural development of the original plan.

1890 Reavell Scott engine.

While the new buildings were being put up, one of the cottages on the opposite side of the Ranelagh Road was taken as a temporary office; E. Bruce Ball, later managing director of Glenfield & Kennedy Ltd., was engaged as chief draughtsman with a few assistants to prepare the detail drawings, and a foreman pattern-maker and two or three men were set to work in a local joinery works so that patterns could be made and castings bought to be ready when the new works was able to deal with them. No foundry was included in the original plan, all castings being bought in, chiefly from the Tortoise Foundry at Halstead, in Essex.

Early in 1899 the buildings had been completed, J. Brown

These three decided to start a company to produce the "Scott" engine (which was very successful, prior to the advent of the high-speed, forced-lubrication engine) and the Reavell "Quadruplex" compressor. Sufficient capital was found with the help of relatives and friends, and the company was incorporated on June 11th, 1898, as Reavell & Company Ltd. Reavell became managing director, an office he held until his death in 1948, aged 82.

Gaskell took on the duties of financial director and secretary, which he continued to perform until his death in 1931. Scott was too occupied in the affairs of his own company in Norwich to take any considerable part in the management of the new company, but remained a director and always attended its board meetings until his death in 1938.

Arnold F. Hills, director of the Thames Iron Works (owned by Hills' father) joined the board of directors as its first chairman. T. R. Elkington was the founder and editor of the *East Anglian Daily Times*, inventor of machinery for the printing press and another friend of Reavell's.

Various places had been considered as possible sites for the new works and Ipswich was finally selected as the home of the new company. The Great Eastern Railway Company, as it then was, had a suitable site to offer on the right bank of the River Orwell near the railway station, and in 1898 about 2½ acres of land

The Ranelagh Works in 1899.

The Ranelagh Works

Chapter Fifteen

Reavell & Co

The three principals of the original Reavell company were William Reavell, his brother-in-law, William Harding Scott, and Charles Gaskell. C. Wilson, a friend and colleague of Scott's, and T. R. Elkington, completed the original board of directors.

William Harding Scott was Reavell's brother-in-law and one of the pioneers of electrical engineering. In 1883 he founded his own company, Paris & Scott in Norwich; amongst his various inventions he had designed a new type of high-speed steam engine specially adapted for direct coupling to electric generators. This was not very suitable for production in his electrical works, so that he was looking for someone to build it for him. William Reavell had an idea for an air compressor which promised to be more efficient and more compact than anything of the kind which existed at the time.

There were then no firms in this country which were giving very much attention to the design of air compressors, which were slow-running machines, mostly horizontal, made by pump makers or general engineers (for example Holman Brothers), and with the increased use of rock drills and other pneumatic tools, it looked as if there should be a growing demand for efficient compressors to operate them.

About the same time Mr. Reavell met and made friends with Charles Gaskell, who had been working in Buenos Aires for a shipping firm. He had recently returned to London and was looking for something interesting in which he could work and invest.

by South Africa and Brazil. In 1951 goods were shipped to more than 90 countries.

Much of the success of overseas sales was due to the policy of sending out factory-trained sales engineers, either on a temporary basis or as residents. BroomWade directors also made many overseas visits. In 1950 Harry Broom travelled 35,000 miles. By 1951 sales of portable compressors were well into four figures, with half the output going overseas. Parallels can be seen here with Holman Brothers and the efforts of the directors in their travels overseas.

In 1958 a merger with Holman Brothers was mooted, to create a British-based organisation with the resources and product range to compete effectively in world markets. Negotiations were called off by Harry Broom. Possibly he was ill; certainly he died on September 12th that year, aged 83.

In the early 1960s a piece of land in Hughenden Park was bought to create the BroomWade Sports Club, with cricket, football, tennis, bowls and a clubhouse, a facility widely used by employees and non-employees. Before that, sports facilities bad been leased in Totteridge. The company's first computer, an ICL1902, was installed in 1966 for production control and in 1968 a new three-story design and development centre was built at High Wycombe. It was in this year that the merger with Holman Brothers finally took place. The new company was given the slightly cumbersome name of the International Compressed Air Corporation. Four years later the name was changed to CompAir, the title acquired from the BroomWade apprentices, having been the title of their magazine. They were paid the handsome sum of £50 for the name. Power tool and product finishing production, which had been carried out at High Wycombe since the 1930s, was transferred to a purpose-built factory in Ystalyfera, South Wales, under the new name of BroomWade Power Tools.

factor in the rapid and economical production of aircraft, guns, tanks, fighting vehicles, ships, rail rolling stock, and practically every type of munitions of war.

The demand for compressors continued unabated but production was limited in 1947 by a shortage of cast iron. The company set about updating and mechanising its foundry to more than double its output. Throughout the works machine tools and plant were upgraded. In 1948 the company celebrated 50 years in business. The news bulletin writer paid a tribute to the Chairman and Managing Director -'the Skipper still on the bridge'. The success of the company, he wrote, was the result of his very genuine and lasting concern for the well-being of the Company's employees, as instanced by the Profit Sharing Bonus, and Pension Schemes which he had instituted. The company had expanded and now had branch offices and works-trained district engineers in every region of the UK, subsidiary companies in Australia and South Africa, and nearly 50 accredited BroomWade agents with branch offices and service facilities throughout the world.

In the 1950s BroomWade developed sliding vane portable compressors. A completely new range of road breakers was developed for use with these compressors, the RB440. This was a revolutionary lightweight machine, weighing only 44lb without steel, economical to run, and designed for easy manipulation and freedom from back-kick. The export market had changed by the 1950s. While Australia had long been the biggest export market, India was now second, followed

Modern V type compressors.

1953 advert for BroomWade products.

than 50% between 1937 and 1941. By early 1941 all departments were working six days a week and overtime, with two shifts in some machine shops. Air raid shelters were built close to each workshop and a spotter sited at the highest point above the works operated klaxons to give warnings of raids.

Women fitting out a Churchill tank in 1943.

Later, Churchill tanks were manufactured at the works, and according to the chairman's report for 1946 "We actually offered to do this work without any profit at all, but the offer was declined." Construction of the vehicles eventually took up a large part of the works capacity. Stationary and portable compressors and pneumatic tools were manufactured twenty-four hours per day, and for a long period during the war for seven days a week. These were a vital and essential

railway wagons and stored in heaps alongside the railway siding.

Like many other engineering companies, BroomWade was in a position to take an active part in wartime manufacturing after the start of World War II. In particular, the pneumatic riveting equipment which was already widely used in the aircraft construction industry was now in great demand. Compressed air equipment was used for operating air raid sirens, digging trenches and tunnels for shelters, spraying paint for camouflaging factories, filling sand bags, pumping water for emergency water supplies, and for demolition squads and rescue parries. The Tube Shelter Sanitation Service used compressors to raise sewage up from the Underground to the drainage system. Later, all oil tankers carried two self-contained compressor plants and many were able to keep afloat and reach port even after being badly holed and set on fire by enemy action.

Even with the loss of the European market, the wartime demands enabled the company to expand, with new buildings increasing workshop areas by more

BroomWade workshop in 1936.

company registered BroomWade as its trade mark. When Clay realised that it rhymed with 'British made', equipment bearing the slogan 'Broom Wade, British Made' began to appear.

While the 'Buy British' campaign helped, the slump worsened. Fortunately help came from an unexpected quarter, Russia, which was into a five-year plan for industrialisation and machinery of all kinds was in great demand by its London buying organisation, Arcos Ltd. The City however had no confidence in the Russians, because they had failed to honour some of the financial obligations of the former Czarist government. Harry Broom discovered that Arcos had plenty of compressor orders ready to place. Other engineering firms, he checked, had had their accounts met promptly. He eventually negotiated export insurance, haggled over prices with the Arcos office, and obtained an order for £50,000 worth of plant.

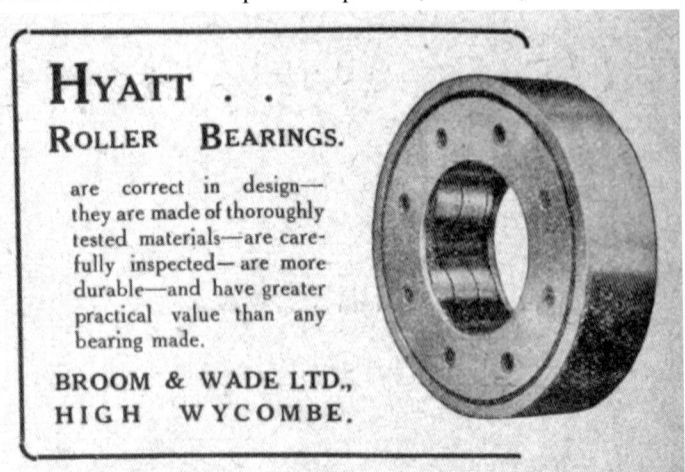

> **HYATT . .**
> **ROLLER BEARINGS.**
>
> are correct in design—they are made of thoroughly tested materials—are carefully inspected—are more durable—and have greater practical value than any bearing made.
>
> **BROOM & WADE LTD.,**
> **HIGH WYCOMBE.**

By 1931 the company was now making reciprocating compressors, sleeve valve compressors, rotary compressors, compressors with mechanical or automatic valves, double acting compressors, single stage and compound machines, steam driven machines and more, each designed for some particular purpose. They were used for anything from filling sausage skins to raising sunken ships. The company survived the slump and in 1935 was floated on the Stock Market. Floating the company raised £48,010 of additional working capital. At the end of the first year as a public company, it showed profits of £25,000. By 1937 trading profit had risen to £65,000 and by 1936 to nearly £100,000.

In 1936 BroomWade acquired BEN Patents Ltd for £100,000 as an entree to the automotive and paint spraying business. They also developed the pneumatic tool and rock drill side of the factory. By 1938 the foundry at the works was producing 50 to 60 tons of cast iron a week. Raw materials were brought onto the site in

Harry Broom's nephew, Christopher Broom Smith, joined the company as an apprentice in 1916. Broom's son Dick joined the company soon afterwards. He later gained an engineering degree and worked for a while with General Motors in the U.S.A. Both later joined the Board, eventually becoming joint Managing Directors. After his father's death, Dick Broom became chairman and, like his father, led an active social life, at one time piloting his own plane and winning yacht races.

In 1928 Harry Broom paid £25,000 for a disused factory at Bellfield, High Wycombe, which had been used for manufacturing aircraft. The building was 500 feet long and included three gantry cranes; the sum included one or two acres of land, leaving room for expansion. Shortly after, B&W bought the business of the Chadwick Patent Bearing Company, which made road breakers, riveting hammers and bearings, and Broom & Wade started designing and developing its own complete range of pneumatic tools. With demand for portable compressors with pneumatic fools growing, a subsidiary company was formed, BroomWade Compressor Hirers, to hire plant.

A major road building programme in Germany in 1929 led to sales of a good number of portable compressors. The company designed a special portable set for the German market. A new design for a sleeve valve type compressor for refrigerators was patented in 1930 and was developed by BroomWade; the sleeve valve compressor was to become the leader in its field for many years.

During the depression the then sales director, Dick Clay, had the idea of fitting a large enamel plate prominently on every portable compressor with a Union Jack and the slogan 'Buy British'. Every tender had a bright label attached, with the Union Jack and the invitation to buy British and save unemployment. It was about this time that the

Broom & Wade's foundry, early 20th century.

war industry. Shortly after his return the factory started to make steam-driven compressors for ship repairs and salvage work as well as 4.5-inch shells for the Naval Command at Dover. The works expanded to cope with this work and more buildings appeared on site. Jethro Wade returned and took charge of the Shell Shop for the period of hostilities. Like many other companies, Broom & Wade had to take on a number of women to make up for the loss of men to the ranks. A few men were retained for heavy work.

The company had problems with the government during the war. Broom & Wade was paid £3 for the work on shells which only cost the company 7s 6d. Refusing to reduce the money paid out, the government subsequently charged Broom & Wade £100,000 of excess profit tax.

By 1910 Broom and Wade were thinking of giving up making woodworking machines, however the final decision seems to have been decided on a whim, when a Mr Glenister bought a machine for £60 less 5%. When payment was asked for another 5% was taken off by the customer for cash. In a fit of pique Broom burned all of the drawings and patterns and decided to concentrate only on air compressors. However the first application for compressors was for paint spraying in the furniture industry.

The Broom & Wade military tractor.

In 1913 the company was thrown into crisis when the man who had loaned the £12,000 asked for it back; the money was not available. Fortunately a former business colleague was able to help out with a bank guarantee, and control of the business was retained. Around this time Broom and Wade separated, Jethro Wade going on to start his own business making deep well pumps, and later valve manufacture, in High Wycombe. Broom did keep the company name however.

Whilst on holiday at the outbreak of the Great War, Harry Broom received a communication from the Admiralty regarding the company's entry into the

General Motors taking over the stock and HSB receiving a share in General Motors in return for the right of General Motors to take over his share at par. Eventually Broom became a director of three companies, Vauxhall, Frigidaire and Delco-Remy-Hyatt with a salary of £1,000 a year. He retained these positions until he resigned at Christmas 1948.

Expansion came with a brick workshop measuring 200 feet by 36 feet and costing £1,700. It contained a gantry crane and had offices at the front. A loan of £12,000 enabled the formation of a private company. It was at this time that the decision was taken for Broom and Wade to design and build a motor wagon. The vehicle had a single cylinder engine which ran on paraffin, and iron tyres. The vehicle's appeal was that it had the lowest cost per ton mile of any type of vehicle, just one penny per mile. The iron tyre design was soon made obsolete by vehicles with rubber tyres. A number of the motor wagons were even sold to Turkey for helping to build a bridge over a pass. In 1908 a number were ordered by a copper mine in Chile to replace their mule-drawn wagons.

1910 Broom & Wade paraffin lorry.

In 1909 the War Office offered a prize of £10,000 for a machine to pull a load of 10 tons over any type of country, plus an order for 50 of the machines. Broom & Wade's paraffin motor wagon competed against 36 others, most of which were steam powered. By the last day of the trials only the Broom & Wade wagon was left to compete against one built by Thornycroft, although the Broom & Wade machine was invariably last back after every trip. Unable to get through a bog, a coil of wire rope and snatch block were used to pull the wagon out. It was the only machine to finish the course but was disqualified because of the tow. The competition was won by Thornycroft and Broom & Wade gave up the motor vehicle business.

Bank Holiday weekends were spent doing repairs at the paper mills, with double pay keeping the books healthy for a while. The salaries at this time were 30 shillings for Wade, 15 shillings for Broom and 45 shillings distributed amongst the other workers.

Broom and Wade believed that the flourishing High Wycombe chair factories, at that time a totally manual industry, were ripe for woodworking machines which Broom & Wade could design and make. Subsequently they sent a circular out to chair shops, factories and paper mills in the area. Unfortunately it was dated Sunday, to which a number of people objected. A Mr Dexter wished them well but suggested that Sunday was not a good day to start and invited them to his chapel.

Later a second-hand foundry, comprising a galvanised building about 50ft by 20ft, with the cupola on an old boiler flue, was acquired. This gave Broom and Wade the capability for casting all of the iron needed for the woodworking machinery and paper mill work. Subsequent work included taking castings for The Hyatt Roller Bearing Company and casting hundreds of pedestals for bearings. Broom eventually acquired the agency for Hyatt's after the company's office in London was wound up. After Hyatt's was taken over by General Motors the agency was lost, but following a visit to the US Broom returned with a new agreement, with

Broom & Wade's foundry, around 1900.

Chapter Fourteen

Broom & Wade

The company of Broom & Wade came about following the meeting of two apprentices, Harry Skeet Broom and Jethro Thomas Wade, at Paxmans in Colchester. The new company was established in 1898, based, as so many others were, on good ideas but little in the way of finance. It was recalled in later years that each partner thought the other had some money; apparently they had £5 and a bag of tools each.

The first company property was a shed measuring 36 feet by 16 feet on a cornfield at Lindsey Avenue, High Wycombe. With no funds to purchase materials, they built it with what they could find. Contacts gave them some timber, cement for the factory floor, galvanising sheets and plenty of nails. An old gas engine, acquired from a scrap pile at the local gas works, was repaired and used in the shed for a number of years. Other scrap piles produced tools while others were made onsite. The office was tiny and the level of illumination given by the gas light varied with the demand on the gas engine, which was used to drive line shafting.

In the early days it was necessary to take almost any jobs around. These included erecting a refrigeration plant at the Brentford Brewery, repairing a butchers' sausage machine and re-hanging the bells on West Wycombe Church, even though it was impossible to ring them because the tower swayed. During this time the company earned a reputation for well-boring, constructing pumps and maintaining water wheels. Experimental work for other engineering companies was also taken on, such as cement spraying for Winget and use of coal gas instead of petrol for Ford cars.

stimated 600 people still waited outside. The proceeds from the concert enabled the Troupe to donate £90 to the Camborne Ambulance. An American visitor to a concert in Newquay congratulated them and said they would do well in the United States while a local promoter wanted them to tour the Western Counties for several weeks.

The traditional minstrel performance followed a three-act structure. The troupe first danced onto stage then exchanged wisecracks and sang songs. The second part featured a variety of entertainments, including the pun-filled stump speech. The final act consisted of a slapstick musical plantation skit or a send-up of a popular play. Spirituals (known as jubilees) entered the repertoire in the 1870s, marking the first undeniably black music to be used in minstrelsy.

Various stock characters always took the same positions: the genteel interlocutor in the middle, flanked by Tambo and Bones, or Sambo, who served as the endmen or cornermen. The interlocutor acted as a master of ceremonies and as a dignified, if pompous, straight man while the endmen exchanged jokes and performed a variety of humorous songs, the beginning of modern stand-up comedy

The minstrels lost popularity during World War II, even though this type of show was still popular in the 1960s, most notably the BBC's Black and White Minstrel Show. This show started in 1958 as a one-off show called the Television Minstrels until 1978.

It is not known how long the No. 3 Works Minstrel troop lasted, nor when its final performance was, however there was a brief encore in the early 1970s when several Holman employees appeared in the Troon Minstrel Troop. These were Roy Thomas (PT Tester), Roy Thomas (R&D), Bill Rashleigh (Packing Shed), Carol Dunstan (Production Control) and Walter Cock (PT Test). Another Holman employee, Paul Rowe (Drill Rig Section), worked on the electrical equipment and sound effects.

BACK ROW: J. Major (drums), S. Rowe (cornet soloist), R. T. Beer (Musical Director), R. T. Knight (violinist), G. C. Uren (banjoist)

FRONT ROW: H. Magor (Tambo, elocutionist, farmyard impersonator), T. Brewer (soloist and Hon. Sec.), J. Jory (dancer), W. Jackson (Massa Johnson), E. Richards (soloist and dancer), E. J. Verran (bass soloist), E. Laity (Bones), S. Keast (Sambo and stage manager)

choir has been privileged to appear with many outstanding artists including Wendy Eathorne, John Treleaven, Carlo Curley, as well as the Bournemouth Sinfonietta the Band of the Grenadier Guards and Black Dyke Mills Band.

In keeping with their position as one of the premier choirs in Cornwall and the West Country, the choir has completed successful tours of America and Europe and is continuing this policy both to gain international experience and to further the good name of Cornish singing.

From 1940 until 1997 the choir rehearsed in the canteen at the Climax works however it became necessary to find alternative accommodation when the canteen was demolished. It now rehearses at Camborne Community School on Friday nights at 7.30 pm.

The No. 3 Works Minstrel Troupe
In 1919 the No. 3 Works Minstrel Troupe was formed from volunteers from the workshop and also included the Holman Quartet. The Troupe proved very popular and gave concerts between St Just and St Austell. The Troupe's third performance was at St George's Hall, later the Scala. An hour before the performance there was a queue down the road; after all of the seats were filled it was necessary to ask the Police for permission to allow another 200 people to stand inside while an

BACK ROW: *J. Major (drums), W. Davey (flute), S. Rowe (cornet), R. T. Beer (musical director), R. Knight (violin), G. Uren (banjo), S. H. Oates (banjo)*
FRONT ROW: *H. Mager (Tambo), G. Chinn (farmyard impression), T. Brewer (soloist), J. Phillips (soloist), W. Jackson (Massa Johnson), E. Richards (Dixie Ned), J. Verran (soloist), E. Laity (Bones) G. S. Keast (Sambo)*

The choir in the Wesley Chapel, Camborne, in 2011.

Benjamin Luxon CBE and Alan Opie, both internationally known baritone singers. The choir has continued with its policy of charitable work bringing help to many hundreds of charitable organisations and thousands of pounds for their various causes.

Over the years, the choir has competed successfully in contests throughout Britain from Llangollen to London. Other achievements include international honours in the BBC's 'Let the People Sing' competition including the cherished Rose Bowl for the most outstanding entry from the British Isles. They have also received awards from the Cornish Gorsedd and have won the Edgar Kessell Gold Cup for the best performance of any choir at the Cornwall Music Festival among their honours.

The most recent success was in May 2003 at the International Male Voice Choir Competition held at the Hall for Cornwall, Truro, when it came overall third behind choirs from Hungary and Finland. This was under the leadership of Roger Wills. Holman-Climax was awarded the shield for the Best Cornish Choir. The

Edgar Kessell receiving the Viscount Nuffield Challenge Shield from Mrs W. E. A. Cullom in 1965.

the radio. The choir received sponsorship from Holman Brothers for years and it only ceased with the final demise of the firm. Later the name CompAir was added to the choir badge to reflect further changes in the company.

Edgar Kessell retired in 1972 after 32 years in charge. He was awarded the MBE for his services to Cornish choral music in the New Year's Honours List in 1975, and a local musical honour, the Lady Trefusis Medal, in 1976. Edgar Kessell died in 1981, aged 61. Edgar's place was taken by George Smith who successfully directed the choir for eighteen years until his retirement in 1990, due to ill health. Roger Wills, under whose leadership the choir continued to grow, directed the choir from 1990 until 2008 when he retired to follow other musical interests.

The current Musical Director is Angela Renshaw, only the fourth MD since 1940. Angela is known for her work to encourage boys' singing in the county. Angela is Musical Director of Cornwall Boys Choir, Boys Treble Choir (a Music Cornwall Ensemble), Cornwall Cambiata and Cornwall Junior Choir, one of Truro Cathedral's four choirs. She is also Programme Leader for Cornwall's 'Sing to Success' singing development programme, and a teacher at Truro Preparatory School, where she is in charge of curriculum music as well as the schools choirs.

In her second year with the choir the current musical director swept the board at the 100th Cornwall Music Festival, further cementing the choir's reputation for outstanding performance. The choir has toured in Britain, Europe and the U.S.A. and it was in America that its camaraderie and high standards led to the town of Calumet, Michigan, becoming a "sister city" with Camborne.

The choir's President was, until his death in 2011, the internationally known Cornish composer Goff Richards and has as its Honorary Vice Presidents

(Holman Group) although the (Holman Group) was invariably omitted, even in the Holman Group Magazines.. A number of the old Holman choir members then joined this thriving group of singers and in 1960 the choir totalled 40 members. In that year the choir visited London to sing for the London Cornish Association and compete in the London Musical Competition, which it won.

In 1961 the choir came of age and performed a 21st Anniversary Celebrity Concert in the Climax works canteen on Sunday April 16th 1961. At the event Kessell was presented with a silver salver in recognition of his distinguished services in attaining a remarkably high standard of performance enabling the choir to achieve high honours in music festivals, the Cornish Gorsedd Shield (1960) and participation in BBC programmes thirty five times. In 1965 the choir was down to three original members (A. Skewes, A. R. Dunstan and Edgar Kessell) but could still put out forty members. In that year the choir won the Viscount Nuffield Challenge Shield at Oxford, for the second year running, and also won the W. Davis Cup in the Championship Class.

By 1965 it became Holman-Climax Male Voice Choir "to mark the long and happy association with the company since the link was established when Climax was taken over by Holman Brothers". In that year it broadcast sixty times on

The Holman-Climax Choir at Rosewarne in 1965.

Climax Male Choir 'Coming-of Age' 1940-1961 outside the former Climax building.

Back row: B. Marks, T. Williams, A. Williams, D. Mitchell, K. Berryman, H. Moyle, L. Williams, K. Bowden, H. Manley, D. Curtis, J. Bowden, N. Truscott, G. Martin

Middle row: A. Knuckey, S. Mitchell, H. Teague, G. Skewes, B. Bowles, T. Kessell, G. Curtiss, H. Berryman, R. S. Dunstan, D. Kessell, J. Uren, K. Notley, G. Bray, R. Bowden

Front row: G. Hendy, A. R. Dunstan, L. Campion, M. Williams, R. Beckley, W. G. Bennett, E. S. Kessell, A. Skewis, C. Glasson, N. Finn, O. Thomas, C. Roberts

The Climax Orchestra at Falmouth Pavilions in 1930.

at the Redruth Ambulance Headquarters in February 1941 with Edgar Kessell as conductor.

At this time, possibly even earlier, a light orchestra was formed, with Jimmy Thomas as Conductor; working together the orchestra and choir organised four successive series of celebrity concerts. Even though these were held during the war-time winters the concerts were sell-outs and acquired £6,000 for war and other charities. At the invitation of the BBC the choir broadcast nine times over the radio.

After the war Edgar Kessell continued as the Musical Director, and over the years built up the small group into a well-respected and well-known choir. Edgar had also formed the Treverva Male Voice Choir (in 1936) and Mabe Ladies Choir (in 1940).

Amalgamation
Following the take-over of Climax by Holman Brothers the choir representatives in 1958 were L. J. Williams at Climax and R. D. Dunstan at Holmans. During this period the process of changing the choir's name to recognise the reorganisation of the companies commenced, and in 1962 it became known as The Climax Choir

The Holman Quartet.

In 1954 the Choir sang at the funeral of Kenneth Holman, though by this time the choir was in decline, and stopped singing shortly after.

The Climax Choir

Two men from the Climax Rock Drill and Engineering Works, Jan Curry and Trevor Bray, saw the air raid fires in Plymouth while fire-watching one night in 1940. They wanted to do something to generate funds in order to alleviate distress in the city, as well as London, and decided to form a small choir; this became the Climax Male Voice Choir.

This choir went on to become one of the premier choirs in the country, raising many thousands for charity and achieving high honour in festivals throughout the land. The first concert was quite humble and took place

The 1941 Climax Orpheus Choir.

Jim Holman receiving the Queen's Award from the Lord Lieutenant of Cornwall.

was presented by the Lord Lieutenant, Sir John Carew Pole, in the works canteen on October 6th. This was followed by a visit from the Queen Mother in May the following year when she was received by Mr Jim Holman, chairman of the company.

During her visit, the Queen Mother unveiled a commemorative plaque by breathing across a fluid logic system switch which, much to her delight, caused the curtains to part. So impressed, she asked for a repeat performance.

New machines were still being added, like the general purpose and very small Silver Feather and the Silver 82, which was developed from the ubiquitous Silver Three.

One feature of the centralisation scheme was that the old house of Rosewarne, with its grounds, was no longer required once the new admin building was finished. John Holman took the view that the house should "continue to have a family association of some kind rather than become an impersonal institution."

Bill Goldsworthy (above) and Dick Jenkin at the Test Mine.

After considering a number of options it was decided that Rosewarne would become the Gladys Holman Home for Spastics, bearing in mind her work in hospitals in the district. At this time the term "spastic" was used to describe cerebral palsy.

In 1996 the house was sold by Scope to a Truro-based developer for conversion to luxury flats. However, since then it has been standing neglected and boarded up, much to local concern..The Grade II listed building, which dates from 1815, is now on sale (in February 2011) for £250,000. The asking price is low because of the amount of money required to restore the Georgian property.

In recognition of the company's outstanding export performance it was awarded the Queen's Award to Industry in 1966 while a second Queen's Award was received the following year, this time for technological innovation. The award

The new admin building in 1965.

new roofs built on the site and new sewers had to be laid. This was a very complicated job, starting with locating a stream at the end of Dolcoath Road which discharged into the Tuckingmill Valley. A gravity drain was then laid from the PCA building in Foundry Road through the Stationary Compressor Test Shop; it then picked up the discharge from the S & A (Stores and Assembly) roof and, eventually, discharged into a 'wet well' constructed at the end of the garden of No. 2 Dolcoath Road, which had been bought by the company for that purpose. A pump house was constructed in the garden, now just inside the new site entrance, to house a battery of three pumps; these could work singly or in unison, depending on the demand. The pumps forced the water through a 9-inch bore into an inspection pit just outside the entrance. From here it went to a 24-inch gravity drain which also carried the water from Plastics & Maxam, R & D and the new admin building.

In 1965 the Holman SV 40 rock drill was introduced. This durable drill featured a standard air cock instead of an air control with integral back head and a plain bolt with extension arms instead of the usual 'D' grip. Weighing 40lbs the SV 40 could drill twelve inches per minute into Cornish granite at 2,375 blows per minute.

number of projects could 'proceed peacefully' side by side.

The centralisation scheme was hailed as a great success by the company in the autumn of 1965. The scheme had involved the re-siting of three different works, the movement of 650 machine tools, the installation of several miles of pipe-work and construction of several new buildings, all within 21 months. The extension to the balcony stores in No. 1 Works Heavy Machine Shop had to be structurally completed within the annual holiday of 1963. Preparatory work involved alteration to the goods lift, setting out, excavation and concreting of stanchion bases and the re-arrangement of approximately 40 machine tools, all carried out during the four weeks prior to holiday. A new compressor assembly building had been erected next to the No. 1 heavy machine shop, with work transferred from Dudnance Lane. Holman masons and electricians had been involved in work, the main part of which was to be carried out by Mowlem.

The new works caused concerns regarding storm water because the existing sewers were incapable of handling the water produced by the large area of

An early part of the centralisation project: the track from the Roskear branch is being dug up and work has started on the new stores building. PT is on the right, behind the old platform.

J. F. (Mr Jim) Holman.

Station. To the east of PT the area immediately north of the Central Steel Stores was levelled for the construction of two extra bays for a grinding shop and a tool room while to the east a new Research and Development building (26,300 square feet floor space) was constructed to the north of the old tool shop. R & D was being carried out at the old Climax works at Pool until this time. A new four- and five-storey admin building was constructed on the site of the sports field, sporting activities having been transferred to the social club at Roskear.

The 55,000 square foot admin building was finally occupied towards the end of 1965. The 'topping-out' ceremony (tamping down the last part of the concrete on the roof) was performed by James Holman, Chairman and MD of company. Some of the earliest staff were from the Central Design section who had previously moved from the Carn Brea offices to the Maxam/Plastics building. The R & D building was completed at about the same time (this department was also transferred from Carn Brea) and this building included a number of semi-soundproofed test and development test cells so that a

Jill Gundry and Mary Lawrence in the secretarial department. What looks like part of a lawnmower is the telex machine.

253

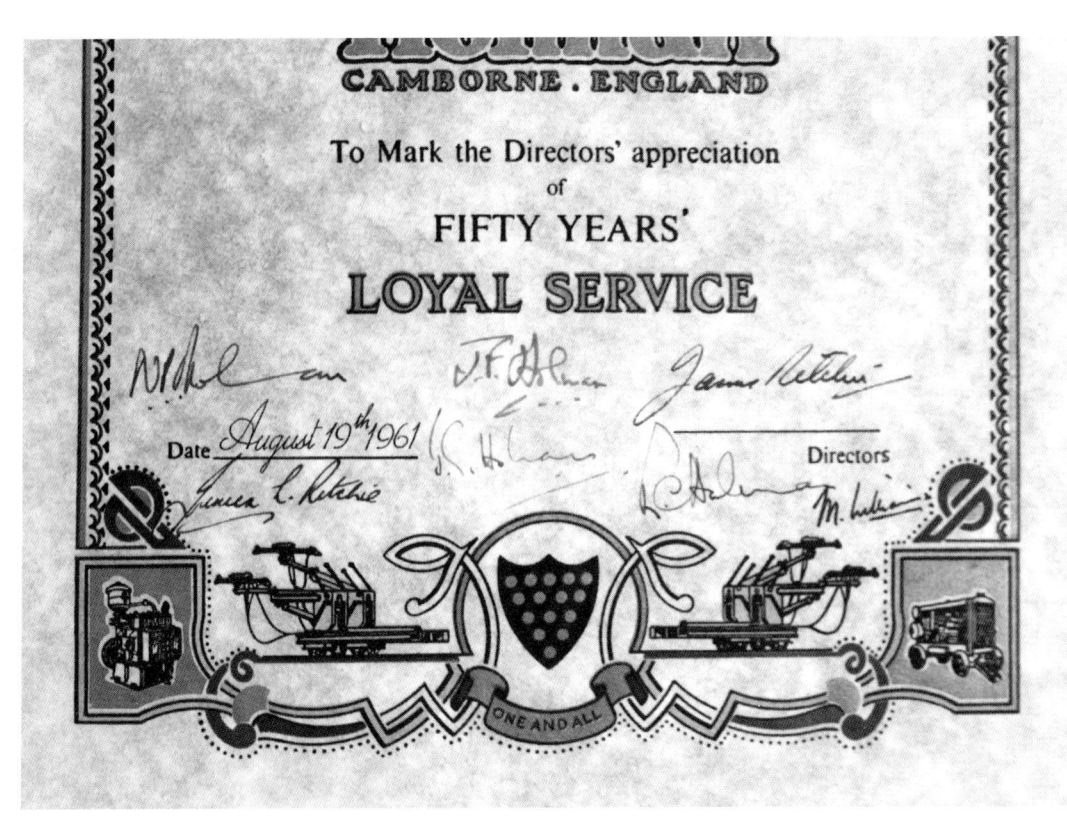

Unlike other employees, Percy's certificate was signed by all of the directors.

the Board, his place being taken by James F. ('Mr Jim') Holman. This news was actually circulated in 1960 – perhaps Mr Percy changed his mind! From September that year James Ritchie was appointed Managing Director jointly with James Holman.

The works was still expanding and in the early 1960s a centralisation scheme was put forward. The earliest work commenced in 1963 when a row of fifteen cottages in Centenary Row East was demolished to make way for a new Portable Compressor Assembly (PCA) building. Prior to this, compressor assembly had taken place at the works in Dudnance Lane. An extension to the balcony stores of the No. 1 Works Heavy Machine Shop was also carried out at this time.

The main phase of work commenced in 1964 with the clearance of some of the buildings to the south of PT to make room for a new three-bay building for the production of Maxam valves and fibre glass compressor covers; production of the latter had been switched from the No. 3 Works opposite Camborne

Percy, almost smiling, receiving his 50-year Long Service certificate from the smiling Jim Holman; the other directors look on.

poster bed that, despite coming to Holmans second hand in 1919, regularly and accurately milled the slides on the big SL.14 drifters, which one old timer remembered being sent to the Witwatersrand in batches of 25 a week. Then there was the famous batch of drifters for Finland which were dismounted from their cradles and fitted with straight bar handles like a road ripper, which led to endless speculation that they were going to be used by some race of Nordic giants.

In 1961 the directors secured the use of a room in the Camborne Public Free Library (the Camborne Literary Institute in Chapel Street) for the free use of the company's pensioners, now numbering over 200. The idea behind this was to form an association for maintaining contact between the company and its former employees. The room had also been furnished at the directors', more properly the company's, expense and was available as a leisure facility. The facilities included rest and reading rooms, with newspapers, periodicals, billiards, and game rooms.

In August 1962 Percy Holman retired as Chairman, although he stayed on

T36D compressor and Road Ripper in use in Singapore.

ventilation louvres in the roof. Nobody could resist the temptation to use the small 'Yale' trucks as high-speed scooters, occasionally with disastrous results. The door on one foreman's small office in PT was invariably left unlatched and, with a smart wallop from a fast moving Yale could

Pendulum milling in the Holbit Works.

easily be flung wide open. That is until the night he chose for once to shut it, and the resultant spectacular impact left him still seated at his desk but hidden by the wrecked door. The quest for a 'bit of stick', firewood to the uninitiated, was an obsession throughout the works and no box or packing case could survive for long, regardless of size.

Clive Carter could recall one exceptionally large and highly expensive packing crate, parked under the full glare of floodlights, that physically shrank as the night shift progressed and that without any of the sounds of the usual surreptitious sawing. By dawn it had most certainly been reduced to kindling, chopped neatly to fit scores of 'croust bags'. Some of those on night shift would be equally tempted to snatch a quick doze, occasionally with unforeseen results. One radial driller, secreted under a pile of bags on top of a locker, found himself being used as a convenient step by his foreman, who, despite being tall found it necessary to jump up and down to reach a window; he was wearing large 'Hobby Boots'.

There were still many old hands who would regale the newcomer with stories of Holmans in the old days and there were other reminders, like copies of drawings for machines initialled by some long forgotten draughtsman back in the 1920s. Or there was the marvellous horizontal milling machine, resembling a vast four

Cornwall, to wit

KNOW ALL MEN BY THESE PRESENTS
that **WHEREAS** _____
residing at _____
occupation _____ born in or hailing from
the County of _____ has expressed his desire
to become a **CORNISHMAN** by process of naturalization
and in support of his application has declared: _____
_____
_____

AND WHEREAS the supplication of the aforesaid applicant has been willingly sponsored by _____
_____ and _____
_____ whose Cornish pedigrees and credentials have been scrutinized and found to be valid

AND WHEREAS the Kernow Antedeluvian Order of "Jan-Lukers" Inc. (headquarters situate in Camborne in the County of Cornwall) has assumed responsibility for vetting such and similar submissions and applications and has signified its approval to the granting of Cornish nationality in regard to the aforesaid _____

IT IS HEREBY DECLARED AND ENACTED that from henceforth the said applicant shall be known, accepted and respected as a member of the Cornish Race, entitled to receive and share all the privileges, advantages and perquisites thereof — both tangible and social (if any) and to such benefits as may arise therefrom, directly or indirectly, real or imaginary.

For and on behalf of
The Kernow Antedeluvian Order of "Jan-Lukers" Inc.

_____ President

_____ Secretary

Signed, sealed and delivered by the Kernow Antedeluvian Order of "Jan-Lukers" Inc., this _____ day of _____
Nineteen Hundred and _____ in the presence of _____
_____
_____

* Where the context permits, masculine embraces the feminine.

Bodmin. 1st July. 1960.

The Editor,

Holman Group Magazine.

Camborne.

Dear Sir,

Cornish Naturalization

It is with deep concern that I read in your Summer issue of the steadily increasing number of people who are applying for Cornish Naturalization, and I feel it is high time that this movement was severely restricted.

The Plague of Locusts which gave Pharoah of old such a headache is as nothing compared with the hosts of foreigners who are daily descending on our Delectable Duchy, taking over our jobs and even our homes.

Men of Cornwall, I charge you, treat these foreigners with the gravest suspicion. Take care that you do not nurse a viper in your bosom. Are your memories so short that you have already forgotten the havoc caused by the infiltration of fifth-columnists during World War II?

I am firmly convinced that some traitor has revealed the secret of my life's work, and that behind this influx lies a deep-rooted scheme to sabotage my own plans for Home Rule for Cornwall.

I know that all true Cornishmen will wish to rally to my side when the time is ripe, so my plans - which include Crofty Miners to demolish the Saltash and new Tamar Bridges, Fishermen from Newlyn and Newquay to lay anchors off the Land's End to prevent us being swept out into the Atlantic after the bridges are blown - will be revealed in detail in your next issue, provided I can give my keeper the slip for long enough to write!

Yours faithfully.

PRO BONO CORNUBIA

The committee was undeterred by "Pro Bono Cornubia" and encouraged more people to apply for naturalisation. The naturalisation movement disappeared from the Magazine after this and it is not known how many successful applicants there were.

At the beginning of 1961 Henry Favell, managing director of Climax Rock Drills (India) and Frank Coombes, Chief Designer of Holman Group had successfully applied for naturalisation. The committee had been joined by C. Patheyjohns and the first certificates were issued in July 1961.

PT never had the bleak isolation generated by the ultra modern production lines, mainly as the work itself came in what were, compared with a car plant, ridiculously small batches. There was still the atmosphere akin to that of belonging to the crew on a large ship, where dry, often riotous humour made bearable the long night shifts, or standing before a machine covered with sweat, oil and dirt and seeing nothing of the hot summer's afternoon except a glimpse of blue sky through the

years that I now feel part of it.

Two letters of support for Noel appeared in the Holman Group Magazine for Spring 1960:

Swansea. 5th January. 1960.

TO WHOM IT MAY CONCERN

Maister Noel Viddes

I do understand that ee referred to above, ave applied for papers of naturlazation to allow un to live without fear of molest in the province of Cornwall, in view of his long assoziation with our County, and ave asked for references from two Cousin Jacks who ave knawed un during that time.

I ave knowed un for some bra while now and would be prepared to speak far'n as I ave always found un to be quite sociable, aving ad the oner of servin unner un in thicky 'Ome Gard: coosed the lil white ball wedd'n over Teidy, and had to ave dealings weth un in the matter of work.

I do think that any man oo can stand Redruth Brewery and Treluswell ales for such a time as ee as, whilst still keeping his self free of ulcers after aiting our pasties should maake a alf daicent citizen; and I will vouch for'n.

Although ee dont knaw it this reference'll cost'n a pint.

(signed) JAN HOLMAN.

(who appends as "His mark" a replica of the Cornish Shield bearing 15 bezants.)

73. Wesley St., Camborne.
27.1.60.

Dear Mr. Editor. Sir.

I 'eard that only won person 'as come forth to sponsor Maister Fiddes' naturalization papers, an' I thought thee wooden mine if I let ee naw I ave nawn Maister Fiddes for nigh on 20 year.

I fust met un in the Home Guard; ee was a posh young chap, a furriner from up-country. Trouble was ee cudden talk our lingo, but we didden 'old that aggin un, it wodden his falt his eddication was neglected, but we lernt un some pile an' thas wot maade un the CO of our Platoon.

All the boys dearly lov'en an' would do anything for un, an' we all thought a pile of what he did for the Ospittle. I'd like to sign that paper for un an' maake un a Cousin Jack.

Missus has just told me "to puddes away that pen and paper thee cussn't rite a letter," but thee cust see for thyself when you read this, she dunnaw wot she's talkin about.

We all hope that when Mr. Fiddes has had one of the greatest onners of the British Isles confered upon un he will remain with us for many years to come.

Yours faithfully.
(sd.) A. H. JOERY.

However there was some resistance to the creation of new Cornishmen and the following letter appeared in the Autumn magazine:

Percy Holman, OBE, a fairly typical pose for him.

so incredibly slowly that they alone seemed to have achieved a supreme balance between gravity and momentum.

It was around this time that a new employee, David Penhaligon, joined the company. He had attended Truro School and Cornwall Technical College where he studied mechanical engineering. As he had decided on a career as a Chartered Engineer so started at Holmans as an apprentice on the understanding that he would be given time off to do an OND and an HND at Camborne Technical College. At this time there was no established apprenticeship for technicians, so Penhaligon had craftsman's training instead. This training took five years and started in about 1964.

At the end of 1959 a movement to grant Cornish nationality to deserving applicants began, culminating in an organisation that was called the Royal Antediluvian Order of Jan Lukers. A committee to oversee applications was formed, comprising Charles Trebilcock, Pat Selwood and Bill Symons (former President of Camborne Old Cornwall Society). A naturalisation certificate was to be given to successful applicants. The first official application was from Noel Fiddes, who was reported to have written

I have dug, mowed, raked and attempted to till so much Cornish soil in the last few

Iberica, S. A., was opened at Pinto, Spain. This operated along very similar lines to Holmans Camborne factory, helped by Camborne personnel. Shortly after the opening it became necessary to obtain a permanent site for testing rock drills, another Test Mine. A suitable site was eventually found about 40km northwest of Madrid at Alpedrete. This became the Holman-Iberica Experimental Mine.

The 1960 Queen's New Years Honours List announced that Mr Percy had been awarded the OBE for his service to industry, in particular to the West of England with its significant contribution to UK and overseas trade development. Despite this, the photograph of Mr Percy that appeared in the Holman Group Magazine was the usual stern study with prominent pipe and no sign of a smile.

By the sixties, the Holman works had developed into a modern engineering works and in 1963 still employed over 2,000. These were still the years of the 'noonday stampede' when, as the hooter blew, hundreds of 'grimy little men in boiler suits and big boots' poured out of the main gate at No. 1 overwhelming anything or anybody that was unwise enough to stand between them and their dinner. There was no such urgency when the hooter blew at one o'clock, and one of the most remarkable sights was that of the many 'old hands' who pedalled their bicycles

In 1959 the company lost a rather small empoyee - Peter the cat, the Rosewarne mouser. He was replaced by Charm, a young Alsation.

Foremen and machine operators, 1958.

really marked the end of the old Holman Empire, leaving Mr Percy Holman as the only survivor of the last generation of Holman boys. Treve Holman's career had been remarkable, embracing a staggering variety of interests combined with what he regarded as his civic, commercial and social responsibilities, while his 'pioneering spirit and original mind' had ensured him a place among the greats of Cornish engineering. He designed a new bathroom for Tregenna as easily as his own new car in 1913 while his extraordinary energy earned him an appropriate epitaph, written for one of the Holman Concert parties during the 1930s:

'When Mr Treve is on the job
the matter calls for haste
He likes the paper on the wall
before you mix the paste'.

In September 1959 a new factory, Holman

Treve Holman

American Newcomen Society in Pennsylvania to receive a generous donation for the Cornish Engines Preservation Society. Considering his age and health, such a trip must have taken a far heavier toll than most realised, yet he resumed his normal hectic activities.

On February 26th 1958 the death of Ralph Ewing, Chairman of the Climax Rock Drill and Engineering Works, was recorded. Ewing was a Scot who had joined the technical sales staff at Climax in 1913. In 1937 he became a director of Climax and in 1950 he became Chairman on the death of his brother Alfred, who had held the post until that time. Ewing continued in the post after the acquisition of Climax by Holman Brothers, with Mr Percy working under him as Managing Director. Another death that year, on May 10th, was that of W. David Rule of the Cardiff Office. Rule was another long-standing employee of the company, having started as an apprentice in 1901. He became the first UK Provincial Sales Representative in 1921 after having spent two years as Assistant Works Manager.

On October 29th Treve gave up his position of Chairman and Joint Managing Director through ill health, the position being taken up by James Holman. He remained on the Board however and retained his position of Chairman. His death, on 6th June 1959, while combining a holiday with business in Lisbon,

Electricians group photo, 1958.

the concept designs for the T.100 turbo compressor came in February 1954. The T.100 was largely built at Holmans with outside firms supplying the specialist jet engine components. Assembly and testing were done in the old Climax stores and canteen, with the T.100 being tested in limited form in April 1958. 1,000

Celebrating Christmas in 1957.

hours of testing and modification ensued when it was discovered the compressor could be started with a 3-speed bicycle. By July 1960 the T.100 was operational but the innovative design was cancelled. Holmans did not have the capacity to make the more advanced gas turbine parts and the fuel consumption, as had always been emphasised, was very high. In the meantime Holmans had developed the first of the 'Rotair' range of single stage, rotary screw compressors that were to become a standard machine.

Holman Brothers Ltd had undergone the first of many changes when, in July 1957, it was announced that, with expansion and post war inflation having 'combined to put a very severe strain on our resources', Holmans was 'now a Public Company' and being quoted on the London Stock Exchange. It was the beginning of a new era and the directors responded with their usual worldwide campaigns for business. Mr Percy vanished behind the Iron Curtain on a month's visit to Poland. Meanwhile Treve Holman, despite increasingly poor health, went 'Round the World in 35 Days', as the trip was headlined in the *Holman Group Magazine* in the spring of 1955. He visited New York, San Francisco, Sydney, Melbourne and Calcutta, still finding time to visit the New York Botanical gardens, though he was thwarted from seeing their collection of rare magnolias, his great love. He also managed to call on the Vice President of the

Another innovative project was the gas turbine air compressor which was initiated after a chance conversation at the 1953 Farnborough Air Show between a Holman director and a designer from Power Jets (R & D). Power Jets was founded in January 1936 by Sir Frank Whittle and others. The second engine built by the company, the W1, powered the Gloucester-Whittle E28/39, the first jet aircraft to fly in the United Kingdom. The W1 was also the first jet engine built in the United States where, as the General Electric I-A, it powered the Bell P-59A Airacomet. In March 1944 the company was nationalised, for £135,000, becoming Power Jets (Research and Development) Ltd. After the Second World War the company was merged with the Turbine Division of the Royal Aircraft Establishment (RAE) at Farnborough, to form the National Gas Turbine Establishment (NGTE Pyestock). The director of Power Jets was one James Hodge. Jim was a Cornishman, educated in St Austell and at St John's College, Cambridge. His experience also included periods at Harwell and as visiting professor of mechanical engineering at Columbia University in New York. He returned to Cornwall in 1958 to join Holman Brothers as chief engineer and was later engineering director, retiring from the board in 1974. A former chairman of the Trevithick Society, Jim died in 2008.

TD20 portable compressor mounted on a North Thames Gas lorry, 1954.

The potential market for this amalgamation was the 'DEW' line, a chain of early warning radar stations to be erected across northern Canada and the goal was a compressor light enough to be airlifted into the wilds and able to stand arctic conditions. Treve Holman became involved and after further meetings,

C. F. E. Vivian in the Johannesburg office, looking very cheerful for an official photo.

Maxam valves, with a 6-inch bore, was made up in haste from spare material for the Greensplat Cornish pumping engine which had been re-erected at a local industrial museum. Part of the production process was the bonding of the piston to the rod, using a special adhesive, which then had to be baked in considerable heat. This tempted one worker to heat up his pasty in the same oven, forgetting that his meal was in a plastic box; the pasty emerged with the crust converted to a form of plastic armour and he was left to scrape out the contents. One development from Maxam was the Fluid Logic Control System which relied upon differential pressures created by the interruption of a constant flow of air.

Another successful adopted product was the Goodyear Pump, patented by James Wallace Goodyear. Working on the Archimedean screw principle, the pump was a superior example of technology that required very fine production methods. Within its limits the pump, although designed for water, was capable of shifting such diverse materials as sand, chemicals and molasses. Many thousands were built until the Goodyear plant was phased out in the mid-1970s.

Dick Gilbert BEM.

The Story of Integration

This is the story of integration
Between two firms who make
Products of a similar nature
And each other used to hate

There was a time in days gone by
When rivalry was strong
When each, to outdo the other
Would vote the other wrong

But Oh, those days have long since passed
And peace has been declared
And now for genuine co-ordination
We're "one and all" prepared

"I wonder where I am to go?"
Was on everybody's lips
Though there was no need to worry
With plenty giving tips

Rumours were running riot
Spreading tales of woe
And tiny bits of paper
Passing to and fro

Now everybody's busy
Getting settled in
Poor "Jan Luke" has "had it"
He's been slipped a Mickey Finn

One versatile product which had been inherited from Climax was 'MAXAM', a sophisticated technology based on fluidics: the use of fluid or gas to actuate switches. The technology had actually been invented in the early days of the space-race, to replace electronic, electro-mechanical and hydraulic devices, which would not work well, or at all, in space. Its uses were wide and various and Maxam valves and allied components controlled all kinds of manufacturing process, from the food industry right through to mining operations. One of the largest

Two photos of the 1951 childrens' Christmas party.

T. Holman and 400 shares to Mr P. M. Holman to enable them to have the necessary holding each to qualify as directors.

By June 1952, Holman Bros. was buying groups of shares (however small) that came onto the market in their own name, but in 1953 the transfer from nominee accounts was in full swing. Almost 678,000 shares traded from March to June, including the purchase of a further 38,000 shares from the executors of Mrs Annie Stephens, William's widow, were transferred at this time.

In October 1957, Climax became a wholly-owned subsidiary of Holman Brothers Ltd. After that date Climax continued to trade under its own name but its activities were increasingly integrated into the Holman group and the relationship had become so close that in August 1962, Holman thought it proper to guarantee the Climax 4½% debenture stock

Unseasonable weather, thought to be the winter of 1946-47.

each was noted, again in the name of nominees, whilst again in September and December a total of 39,000 were added.

At a meeting of Climax directors on the 14th of December 1951 Ralph Ewing, the current chairman, submitted a letter from Percy Holman, dated the 5th of December 1951, stating that Holman Brothers Ltd. had now purchased a majority holding in Climax. The chairman reported that he and Mr Gilchrist had met Percy and John F. Holman at their offices on Monday, 10th of December 1951.

It was agreed to make an announcement that Holman Brothers had acquired a controlling interest in The Climax Rock Drill and Engineering Works Ltd., with a view to utilising the combined productive capacity of the two companies to the best advantage. The notice should be posted on Wednesday 19th of December 1951, and that the notification should be sent to the Stock Exchange, the debenture holders, and representatives and agents both at home and abroad. The company's publicity agents, Messrs Dorland (City) Ltd., should also be notified so that press announcements could be made. Treve and Percy Holman were duly elected directors of the company.

At a meeting of the Holman Brothers board on the 5th of January 1952 the purchase of 100,600 shares at 10s 3d by Morgan Nominees in December was approved and the Secretary reported that as a consequence of Messrs A. T. and P. M. Holman having been appointed to the Board of Directors of the Climax Rock Drill and Engineering Works Ltd., 400 shares in that company had been transferred to Mr A.

Thereupon the Stephens family had agreed to sell 100,000 shares at 10 shillings each, the purchasers in addition paying the vendors' brokers' commission. Mr Grant did not know who the buyers were. This report was considered by the Board and it was agreed that no action could be taken and that it "would be necessary to await developments."

As the market price for the shares at the time was at least 5s 11d and at most 6s 7d, the price negotiated by the Stephens family was extraordinary.

However, from the records, it is apparent that Holman Brothers had been active in this matter for some time. First mention of the purchase of Climax shares occurs in the Holman Brothers minute book at a meeting on the 15th of May 1951 when the purchase of 23,000 shares was made in the name of Midland Bank Nominees Ltd. This was not the first purchase, as it was reported that all the Climax bonus shares accruing to the company's holding had been received by the various nominee companies and that the company's total holding now amounted to 245,498 shares at an average cost of 5s 8d per share.

In August, the purchase in July of 242,354 shares at between 6s 9d and 7s 6d

Climax portable compressor destined for the other side of the Iron Curtain.

A marvellous character study of an unknown employee standing with his back against the radiator of a huge portable compressor. His pictures were used to advertise Climax products as 'Climax Sam'.

Mr Cartwright of Montague Stanley & Co., (his firm of stockbrokers), had been approached by another firm of brokers who asked if the Stephens family would sell a large block of their shares at current market price. The reply given was that they would sell for 10 shillings a share.

Subsequently, the same brokers saw Mr Cartwright and said that their clients would pay 10 shillings per share provided that they could have not less than 80,000 shares and that the transaction should not be marked on the Stock Exchange. Mr Cartwright expressed the view that the proposed purchasers would only be willing to pay this price if the purchase would give them effective control of the company. Mr Grant had reported the position both to the Chairman and Mr Gilchrist. The Stephens' family had asked the proposed purchasers to give some indication of who they were. They had been informed that the purchasers were another commercial firm of British nationality, and that it was their intention that Climax should continue as a separate entity; that they thought that the company could be extended considerably and that this was their reason for making the purchase.

The Climax demonstration van. Date uncertain but not later than 1950 as the building behind is still camouflaged.

Climax office staff, 1940s, possibly during the war, a windy day in Pool.

the immediate post-war years has been chronicled in the excellent publication *Climax Illustrated,* also published by the Trevithick Society.

This period also saw the manufacture of .303 Mk 2 Bren gun barrels with their own unique marking, CRD. (For Climax Rock Drills).

The Acquisition by Holman Brothers in 1951
In the early 1950s, Holman Brothers made a positive decision gradually to purchase shares in the Climax company so that they might add it as the 'Jewel in the Crown' to their rock drill enterprise, in time for the 150th anniversary of the Holman company.

The first mention of a possible take-over occurs in the Climax Minute Book at a meeting held on the 27th of November 1951 when "Mr Grant reported that, as solicitor to the Stephens family, he had been advising them for some time that they should reduce their shareholding in the company in view of the fact that most of their assets consisted of Climax Shares." About two months previously,

World War 2: Above: setting miller cutters at the Bren gun barrel assembly. Below: drilling a gas regulator for a Bren gun barrel.

diameter had been sunk to a depth of 332ft in 31 days by five machines. In May 1926, the earlier record was surpassed at Randfontein, a depth of 386ft being sunk in 31 days.

An interesting view of the construction of the iconic Climax buildings; the exterior wall appears not to have lasted very long.

Further records were claimed in 1929 for the 'Climax Lightweight Jackhammer' of which over 1,100 were in operation in the Rand by 1934. 'The Climax Disc Valve Drifter' was introduced in 1933 and the 'Climax Sleeve Valve Jackhammer' in 1935. In March 1936, the record was extended further to 422 ft. By the installing of 'Climax drill sharpeners', users of rock drills could ensure a constant supply of the properly sharpened steel necessary for the most economic performance.

During these later years, although William was still associated with the company, his influence had waned and he retired to Bournemouth where he died on the 17th of December 1947 leaving an estate of over £45,000. His widow who was in poor health by that time died in 1952.

The Second World War
Much of the contribution of the company during the Second World War and in

Works St John Ambulance nurse, 1930s.

to meet his expenses for a year had been estimated at about £450 and it was agreed that about another £100 be granted for contingencies and minor entertainment expenses. The total expenses to be borne by the company however, must not exceed £600. During these difficult years William spent much of his time in South Africa, even taking his wife and daughters with him. His relationship with the other directors deteriorated to the extent that, by late 1925, Ralph Ewing was authorised to write to him asking him to return by the end of the year.

In his absence, at the meeting on the 26th of February 1926, William's remuneration was discussed and it was agreed that a reduction was absolutely necessary in both his annual salary and director's fee. Ewing was delegated to discuss the matter with William prior to the next meeting. However, at that meeting on the 29th of March 1926, a letter of resignation was received from William and this was accepted, whilst his services were to be retained in an advisory capacity. He was to be paid a retainer of £50 per month on a contract that could be terminated by the board at any time.

Later that spring it was announced that a world record had been set in April at the Randfontein Estates for shaft-sinking by Climax machines. A shaft 23ft in

Opposite: the Climax management in about 1936.

Top: Alfred Ewing (Chairman and Managing Director), R. de H. Saint Stephens (General Manager), Norman E. Rickard (Works Secretary)

Middle: James Gilchrist (Secretary), H. L. Sargent (Director)

Bottom: John H. Crawford (Director), Ralph Ewing (General Sales Manager), N. Baldwin Davies (South Africa Manager)

William Charles Stephens was the other director.

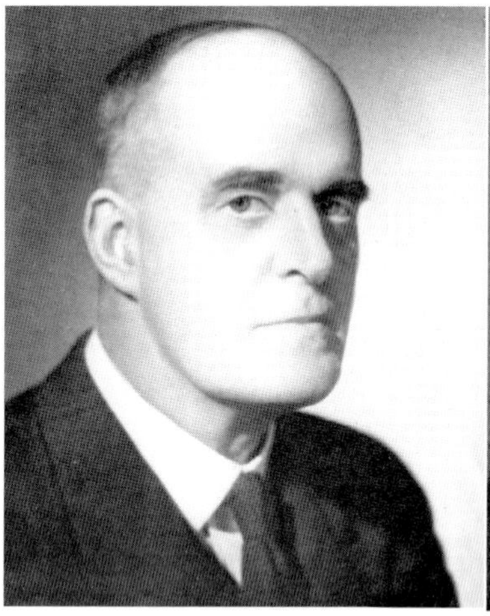
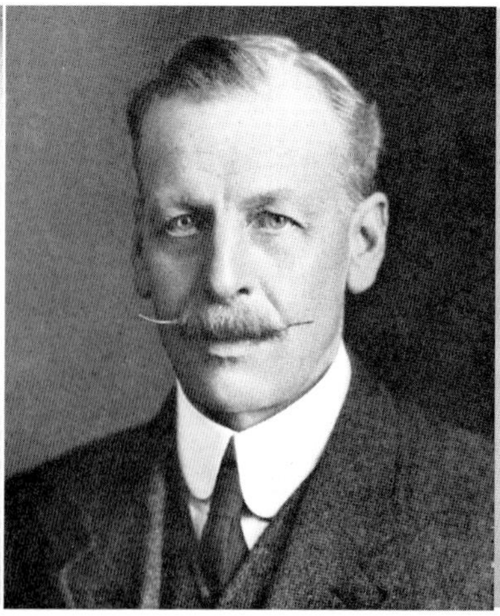
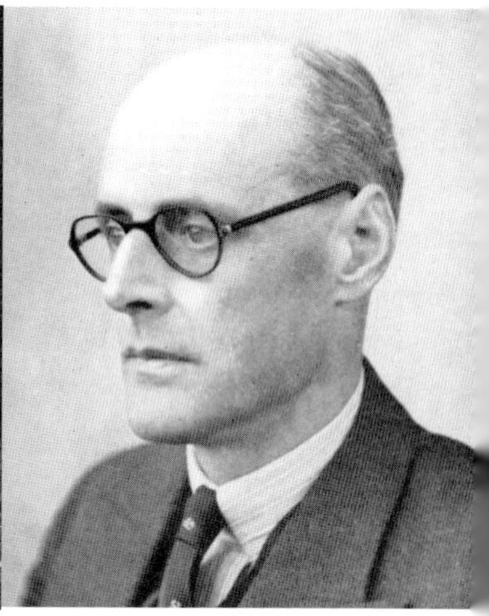
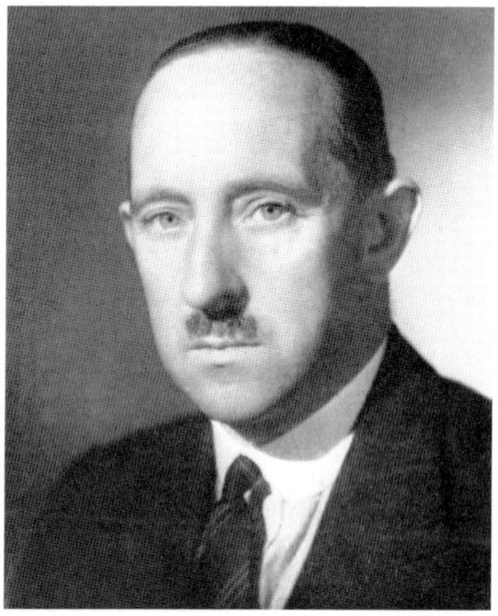
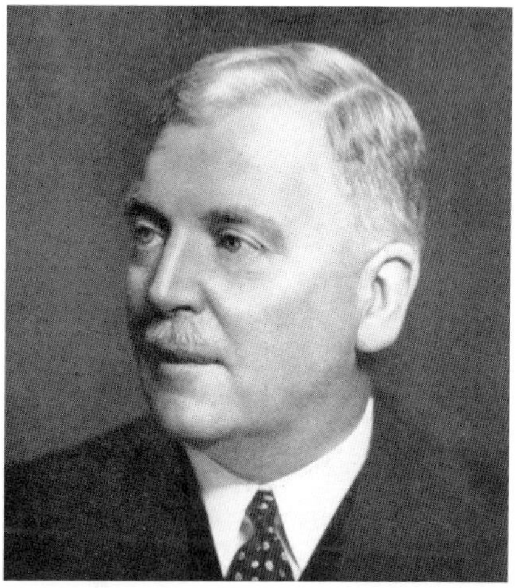
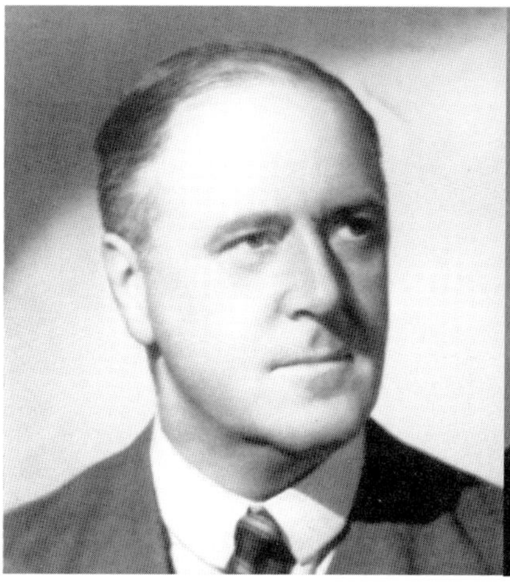
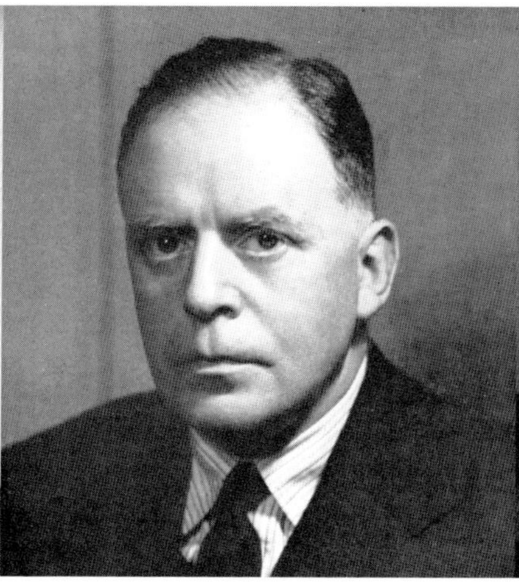
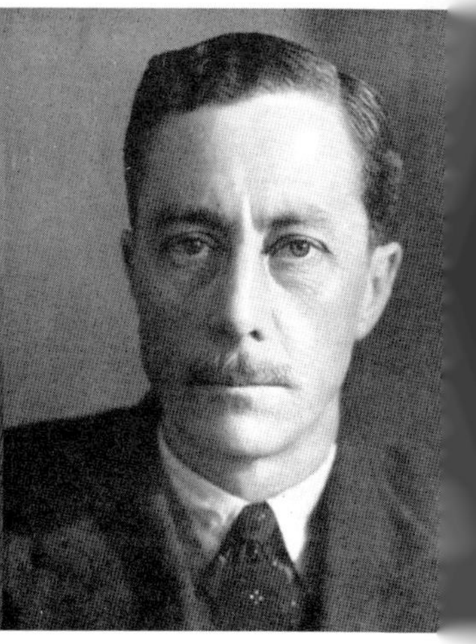

Climax rock drill being used in a granite quarry, 1930s.

by the competition. The other machine was a 4-inch heavy cradle 'Drifter' and had been designed to compete with a similar type of machine manufactured by Ingersoll Rand Co. Ltd. Experiments had been carried out continuously to overcome the piston trouble in the original machine and it was believed that a solution had now been found. Taking these facts into account, Ewing stated that within a short time the company could reasonably expect an increase in sales in South Africa and probably also in Canada.

The necessity of increasing sales if profits were to be made was recognised, and as South Africa was the largest territory for the sale of the company's machines, it was felt that it might be possible to achieve results there in a short time. This would considerably improve the company's position so it was decided that William should proceed to Johannesburg as soon as practicable to increase sales in that market.

However, in view of the financial position of the company, it was resolved that he should be paid only travelling and living expenses, in addition to his present annual salary of £950 and £250 director's fee. The amount set aside

During this period the 'Hydramax' drill, introduced in 1915, was the foremost stoper drill whereas the 'Hydromite' (1917) was a baby stoper. Other trademarks, including 'Aeromax' in 1918, 'Aeromite' in 1918, and 'Aquamax' in 1920 were introduced, but rarely used.

The depression

In 1923, a significant event occurred. William had been issued with a contract at the time of the formation of the company in 1913 under the terms of which he was appointed Managing Director for a period of 10 years. On the 11th of September 1923 the board resolved not to renew this agreement. Instead, he and Alfred Ewing were appointed joint managing directors. As the recession deepened, cuts became necessary and the salaries of the joint managing directors were reduced. The South African business was diminishing and the company threatened to break away from the Rock Drill Convention of South Africa, a trade organisation, should it fail to obtain more business. Meanwhile, William was commissioned to design a new drill specifically for the South African market.

William Charles Stephens in 1930.

By November 1924 the position of the company had deteriorated further and at this time Harold Bendixson was still the chairman. Reviewing the state of trade it was recognised that, because of financial constraints, it would have been difficult to have achieved more. Ralph Ewing stated that two new models had recently been completed and would be available for despatch to Johannesburg and Canada towards the end of the year. One was a 3-inch cradle machine known as 'The Climax Light Drifter' and results obtained on test with this machine showed a very considerable increase in drilling speed over similar machines manufactured

THE CLIMAX HYMN 1917

Our Climax Drills are just the things
For any sort of ground,
Such speed and good economy
Cannot be elsewhere found.

The Vixen Drills we made of old
Can still the records beat;
Unless a Hydromax is there
They fear no drills to meet.

Imperials of any type
Are lighter far, we own;
But that's because there's much more steel
On all their drawings shown.

The new Dl cannot be beat
By any other make,
In mines of coal or iron or tin,
Our customers all state.

Our latest types of Hydromax
Are steel from end to end;
And in whatever place they're tried
They're sure to make a friend.

So anywhere that you can use
This drill your work to do,
You'll get more done,
Use much less air, and cut the costs in two.

Now making shells is all the game,
We keep our end up, just the same.

The Great War
The First World War was a challenging period for the company, as it was for many others. Much of its production was turned over to support the war effort. Foremost amongst this was the contract for shells and production in 1917 was 6,000 per week. Each shell took 28 different machining operations to complete and the finished shell was then painted and varnished. To perform these tasks many women were recruited as munitions workers.

At Christmas 1917, a booklet was published entitled "Crumbs from Cornwall" which was a pastiche on the then popular "Fragments from France". A 64-page document, it recognised in the foreword the harmonious relationship between the management and employees of the Climax works, noting that the employees recognised that their welfare was always the first consideration of management. In several articles it takes a light-hearted look at the company, often poking fun at workers and management alike. In one cartoon 'Mr Will', as he was affectionately known, is shown attacking his pasty with ... what else ... a rock drill! (See drawing on page 217.) It also chronicles the arrival of women in the workforce, whose efforts in shell making were recorded.

An aerial view of the Climax works from 1918. East Hill, Tuckingmill is to the left. Only the building almost opposite the works gates now survives..

A group of workers, nearly all male, at the factory in 1917.

of December 1914, at William's request. George Darlington Simpson remained a shareholder of the company by making private purchases but died on the 28th of August 1919 leaving an estate of some £20,000

A more mixed group of workers, also 1917. The women are most likely working on munitions, the so-called 'shell girls'.

held 21st day August 1914 and that the interim injunction granted by the Vacation Judge preventing the Company, their servants and Agents from acting on the Resolution purporting to remove Mr Darlington Simpson from the office of Director and appointing any other person to fill the purported vacancy.

George Darlington Simpson requested that a similar circular should be sent to shareholders informing them that he had been granted an interim injunction restraining the company.

On the 30th of October, William, Bates, North Lewis (the solicitor) and C. W. Curzon (Secretary pro tem.) were present at a meeting in London when the Secretary reported that the bank had informed him that the company was to consider the facility of the overdraft for £1,000 withdrawn. The matter of renewal would be considered later but, for the time being, the company could only deal with a credit balance.

North Lewis reported that the negotiations were in progress for a settlement of the Action between Darlington Simpson and the company and William, any settlement being subject to the agreement of the directors, but certain points which were considered important were outstanding and until these points had been discussed with Darlington Simpson he would not make any statement with regard to the terms of settlement. To enable North Lewis to meet Darlington Simpson's solicitor and arrange a final settlement before making his report, it was agreed to adjourn until the following day. However, when the meeting was resumed on the Saturday, negotiations for settlement were still not complete and he was therefore unable to make his report. He suggested that the meeting be again adjourned until Tuesday the 3rd of November, 1914.

When the meeting was resumed on that date Lewis reported that the negotiation for a settlement of the action between Darlington Simpson and the company and William had now been completed subject to the approval of the company. The terms had been agreed by British Bank of Northern Commerce, by Darlington Simpson and William Stephens. At the next meeting on the 11th of November 1914 the terms of settlement and a letter of resignation from Darlington Simpson were laid before the meeting and both were accepted. 3,000 Shares were transferred from Darlington Simpson to Harold Bendixson and a further 3,350 from William, paving the way for the appointment of Harold Bendixson as Chairman on the 29th

"Mr Will" at breakfast, as drolly depicted in Crumbs From Cornwall in 1917.

Darlington Simpson opposed the proposals, but William insisted that North Lewis be instructed to communicate them to the bank and to endeavour to persuade them to find the money necessary for the fortnight's pay that was due on the 26th of September. The meeting was adjourned until Thursday the 24th of September and on resumption, North Lewis reported that the bank refused to accept the Company's proposal and that it would not undertake to find the money required for wages.

Another Drill

The meeting was once more adjourned to enable the position of the Company in regard to the funding of wages to be placed before the bank yet again.

Upon resumption it was reported that, under the circumstances, the bank had agreed to find the wages for this forthcoming pay day only. It was agreed that the appointment of Messrs Bendixson and Martin as directors had been invalid. In view of the Court Injunction the Secretary was instructed to write to the Bank, and Lloyds Bank in Camborne in the following terms:-

> Dear Sir, I am instructed by the Directors to inform you that Mr Darlington Simpson commenced an action against the Company on the 1st Day September 1914 for an injunction restraining the Company from carrying into effect the resolution passed at the Extraordinary General Meeting

Darlington Simpson had received in the letter from William. However, this was not the end of the matter.

At a meeting held on the 27th of August 1914 with the British Bank of Northern Commerce it was reported that Darlington Simpson had instructed Martin Stewart (solicitors) to take legal action. They subsequently met on the 23rd of September, with Darlington Simpson and his solicitor, when it was reported that the appeal court had received affidavits from William, Andrew Rutherford, Frank Agar and Darlington Simpson and granted an order "restraining the Climax Rock Drill and Engineering Works Ltd, their servants and agents, from acting upon the resolution purporting to remove Darlington Simpson as director". This restraining order, which was filed at Somerset House, also prohibited the appointment of any other person to fill the purported vacancy and the prevention of Darlington Simpson's attendance from board meetings until judgement was reached.

It would appear, though tantalisingly it was never revealed, that the bank was insisting on Darlington Simpson's removal and that it had been unaware of the assurance given by William in the letter that protected his position in just these circumstances.

Letters from both the bank and Darlington Simpson were read to the meeting and North Lewis advised the directors that under the circumstances they could not possibly agree to all four of the requirements set out in a letter from the bank, the contents of which were again not revealed. Several suggestions were made and the meeting was adjourned until 4:00 pm. to enable William and North Lewis to see the bank on behalf of the Company.

After the adjournment, Lewis reported that the bank refused to depart from the conditions laid down in their letter of the 19th of September. The question was then considered as to whether it might be possible to get the bank to agree to just *some* of the requirements of its letter. It would seem that the bank required two of its nominees to be appointed directors and for the removal of Darlington Simpson but in view of the injunction, the company could not agree to this. As a counter-proposal (and a compromise) the company suggested that any two representatives nominated by the bank should have the right to have notice of and attend board meetings of the company and that such board meetings should be held at the company offices in London.

amendment carried.

William had other ideas, refuted this claim, declared it not carried and then put the original resolution to the meeting. He and Bates voted for this resolution whilst Darlington Simpson voted against. Agar explained that, under the circumstances, he would prefer not to vote. As the resolution was not carried by a sufficient majority, the chairman demanded a poll to be taken immediately.

The votes cast in favour of the resolution were 50,500, with 8,450 against. William, as chairman, declared the resolution carried by more than the necessary three-quarters majority. At this point, Darlington Simpson left, presumably in high dudgeon.

An ordinary meeting of the directors followed, at which William explained that, to meet the requirements of the Bank of Northern Commerce Ltd, he had pleasure in proposing that Harold Bendixson and Alfred Ridley Martin be elected directors of the company.

So it would appear that the move was approved by the bankers despite Darlington Simpson's objection. It was possibly at their insistence that he was removed and perhaps they were unaware of the assurances that

William Charles Stephens, photographed in South Africa in 1892.

his wife, Jane. William now had to make arrangements with his mother to cover this. In 1898 she was entitled to half the business but since then the business had been expanded by William's efforts alone. Clearly, though her share would have increased in value, it represented a smaller proportion of the whole. The settlement she eventually accepted was for 5,000 preference shares that paid an income of 6% pa. but which conferred no voting rights. The Register of Shares reflects these early dealings.

William also assigned leases on the site granted by the Robartes estate to the company, whilst notice was given to terminate a pre-existing royalty agreement with Ingersoll.

William did not take to the strictures of a private company very easily. He was so used to making his own decisions without reference to others that to have other people interfering in his business did not sit comfortably on his shoulders, and within a year those tensions had surfaced.

A Disagreement
At the first AGM held on the 29th of May 1914 things seem cordial enough, with Darlington Simpson proposing a vote of thanks to William for his services to the Company. However, at a meeting on the 13th of August at which only William and Richard Bates were present, they resolved to call an EGM on the 21st of August at Carn Brea to "remove Mr George Darlington Simpson from the office of Director".

The meeting duly took place with the directors and the works secretary, A. J. Crocker present. The Chairman had also invited Messrs Harold Bendixson, Alfred Martin and Milligan, to whose presence Darlington Simpson formally protested, since they were not directors of the company but, in fact, nominees of the bank. Darlington Simpson brought to the notice of the board the undertaking given to him by William that he would not use his own votes to remove him (Darlington Simpson) as a director. He proceeded to read the letter, of which William denied having any recollection, as did Richard Bates, a nominee of the bank.

The resolution to remove Darlington Simpson was then put to the meeting by William as Chairman, but Darlington Simpson moved an amendment aimed at recognising the letter and dissolving the meeting. He put the amendment to the meeting himself and when he voted for and Bates against, he declared the

balance of £9,000 by accepting £2,250 every half year until September 1915. In return they received the £9,000 balance of the debentures as collateral. When the debt was repaid, the debentures were to be lodged with the bank. Arrangements were made to hold monthly board meetings, alternately in London and at Carn Brea, the next meeting to be at Carn Brea on the 7th of October 1913.

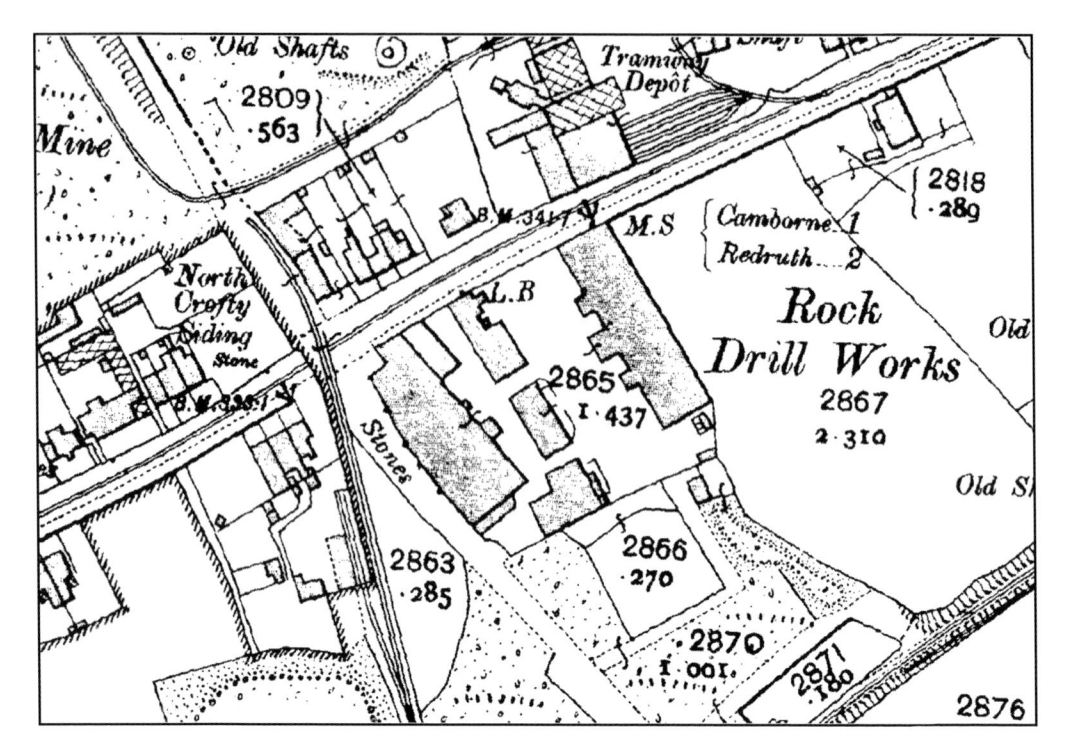

There then followed an event that was to be much debated in the future. George Darlington Simpson produced a letter dated 17th September in which William undertook to exercise his voting power to retain Robert Bates and George Darlington Simpson as directors so long as there was any liability to the British Bank of Northern Commerce. This undertaking was approved and confirmed.

Money was owed to the Capital and Counties Bank which held deeds, documents and patents as collateral. It was agreed to repay this sum and to lodge the patents with the Bank of Northern Commerce but with access on request. The bank also required a personal guarantee by William, backed by three insurance policies totalling £11,500, and that a local bank (Lloyds Bank Ltd., Camborne) was retained.

When Richard Stephens died in 1898 he had left his share of the business to

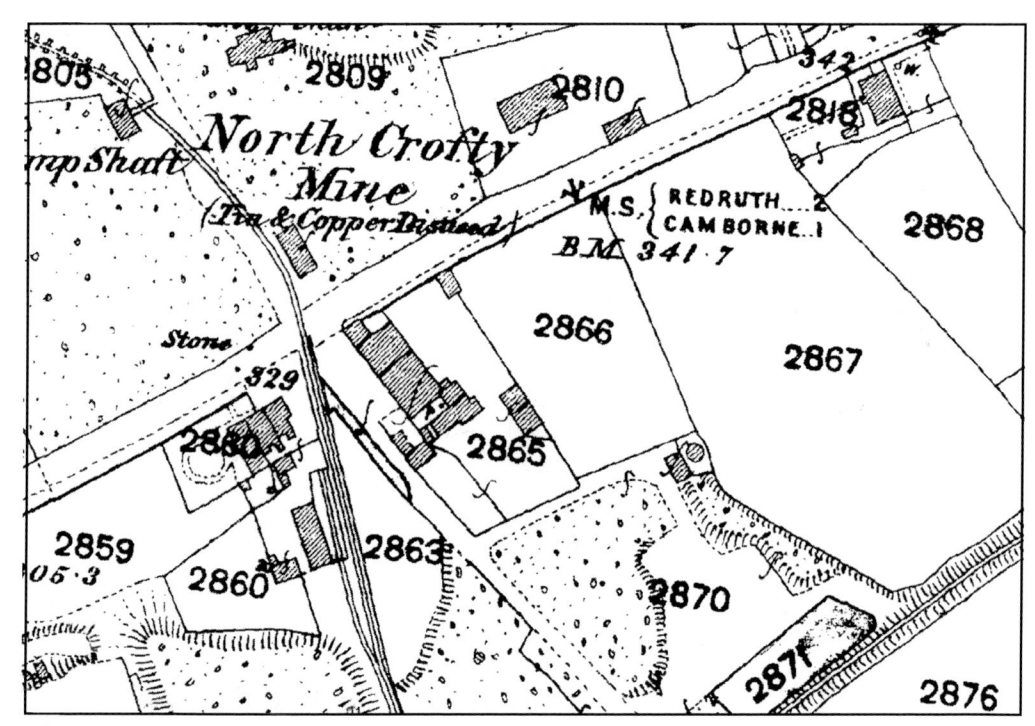

These two maps show the great expansion of the Climax works from 1880 above to 1908, right.

The British Bank of Northern Commerce was appointed bankers, W. B. Peat & Co, auditors and Downing, Handcock, Middleton & Lewis were to be the company's solicitors. Frank H. Agar, a chartered accountant and official nominee of the Globe Trust, was appointed company secretary, whilst Alfred Crocker was made works secretary at Carn Brea. In the minutes it was recorded that two directors were required to form a quorum and, whilst two shares were allotted to the signatories of the Memorandum and Article of Association, the balance of the 60,000 £1 shares was allotted to William.

By agreement, William was appointed Managing Director for a period of 10 years. Debentures to the value of £20,000 bearing 6% interest were created with Sir William Barclay Peat acting as trustee for the Debenture holders – for a nominal fee of £25 pa. £11,000 of debentures was deposited with the Bank as security for a loan of £11,000.

Harris & Co, who were purchasing agents for the company, agreed to accept a cheque for £6,716 13s 11d on the 20th of September and to extend credit for the

because of his knowledge of the business to be conducted. William Charles Stephens was appointed Chairman of the company and George Darlington Simpson and Robert Bates his fellow directors.

Like Holman Brothers, Climax produced several sports teams. These pictures show the 1916-17 football teams, juniors above and seniors below.

arranged that the quantity of water to be discharged could be instantly regulated from a fine vapour to a heavy shower. The device would draw water horizontally from any distance and could be arranged to lift water 25ft and was not a separate device, but instead formed part of the air tap of the drill. The same water, if collected, could be used over and over again and the device did not 'choke' with the use of dirty or gritty water.

In 1906, Holman Brothers were awarded a First Silver Medal for a similar device but by then Climax had moved on and were awarded a First Silver Medal for an adjustable rock drill cradle. In 1910, the Society introduced a new award, the Grand Diploma of Honour, one of which was awarded to R. Stephens & Son and the other to Holman Brothers. By 1914, William Stephens was elected a Vice-President of the Polytechnic Society.

Climax Rock Drill and Engineering Works Ltd.: First Steps
With the continued growth of the business, in 1913 William decided to launch a private company in order to raise capital necessary to expand further. From debenture documents issued in March 1913 for "The Climax Rock Drill and Engineering Works Ltd" we learn just how far William had come since the death of his father in 1898.

There were two directors prior to the debenture offer, William Charles Stephens and William Thomas, the consulting engineer. They held a lease for 35 years on a site of 2.5 acres and Ingersoll Rand paid royalties to produce the Climax Simplex tool holder under licence. The business already extended to South Africa, India, Spain, Australia, South America, Mexico and Great Britain. Profit for the year 1911 was £8,000 and for 1912 it had risen to £12,500. Total assets were over £52,000. Over 53 patents were cited in 11 countries, and 135 mines named as customers.

The first meeting of the directors of the Climax Rock Drill and Engineering Works Ltd. was held at the registered offices of the company at 9–11 Fenchurch Avenue, London on Wednesday the 17th of September 1913 at 2:30 pm. Those present were William Charles Stephens, Robert Bates, George Darlington Simpson, H. North Lewis (from solicitors Downing, Handcock, Middleton & Lewis), and Frank H. Agar, a nominee of the Globe Trust.

George Darlington Simpson was asked to take the chair for the initial meeting

Another logo, this one from 1905.

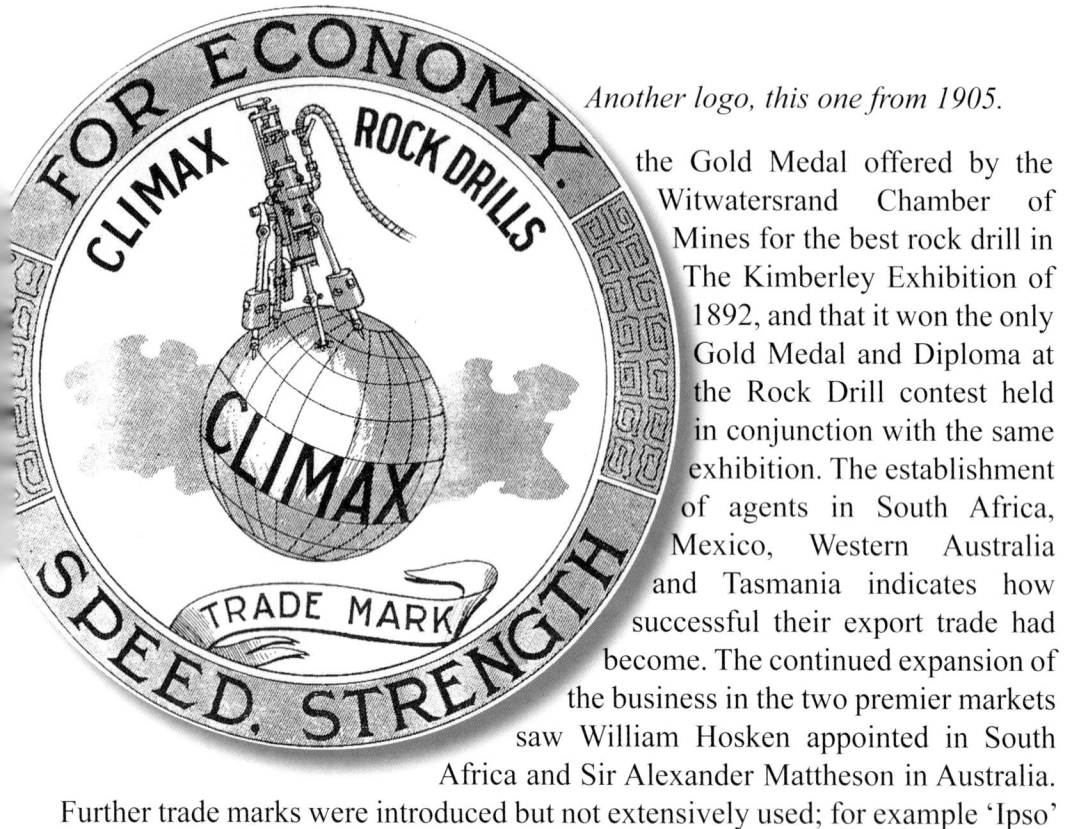

the Gold Medal offered by the Witwatersrand Chamber of Mines for the best rock drill in The Kimberley Exhibition of 1892, and that it won the only Gold Medal and Diploma at the Rock Drill contest held in conjunction with the same exhibition. The establishment of agents in South Africa, Mexico, Western Australia and Tasmania indicates how successful their export trade had become. The continued expansion of the business in the two premier markets saw William Hosken appointed in South Africa and Sir Alexander Mattheson in Australia. Further trade marks were introduced but not extensively used; for example 'Ipso' in 1908 and 'Knock-out' in 1909

Royal Cornwall Polytechnic Society
The Royal Cornwall Polytechnic Society, which had been founded in 1833, was the premier institution showcasing the technological advances in Cornwall.

In 1882, at the Jubilee Exhibition, rock drills designed and manufactured in Cornwall were exhibited for the first time. On this occasion, Messrs McCulloch & Holman were presented with a First Silver medal and R. Stephens & Son with a First Bronze medal. This was the start of an intense rivalry that was to last for many years. A rock drill competition took place on September 13th and 14th of that year with six drills taking part. The results were much debated in *The Mining Journal* with claim and counterclaim from the competitors.

In 1904, R. Stephens & Son was awarded a First Silver medal for their 'Little Vixen' drill with patent 'Climax Dust Allayer'. The report states that the Dust Allayer was intended to obviate or minimize the occurrence of phthisis, which was chiefly due to the inhalation of the fine dust given off during machine drilling operations. It killed all dust from dry holes. The nozzle of the device was so

reliability was due in the main to the fact that the drill had only two moving parts. Its reputation was cemented at South Crofty, which remained loyal to the Climax Company over many years, despite intense pressure from Holman Bros.

A world record, claimed in 1907, was gained at the Juniper Mine in South Africa by an employee, Henry Glasson, using a 'Little Wonder' stoping drill. Henry was later presented with a gold watch by William to mark his achievement. Henry Glasson had been apprenticed to William in 1901 and the watch and indenture of apprenticeship are still in the possession of Henry's grandson.

An ornate invoice from 1908 records proudly that Climax was the winner of

Henry Glasson testing rock drills in South Africa in 1907. He was identified by the chain of his pocket watch (right), which is very distinctive.

New Climax letterhead, 1906.

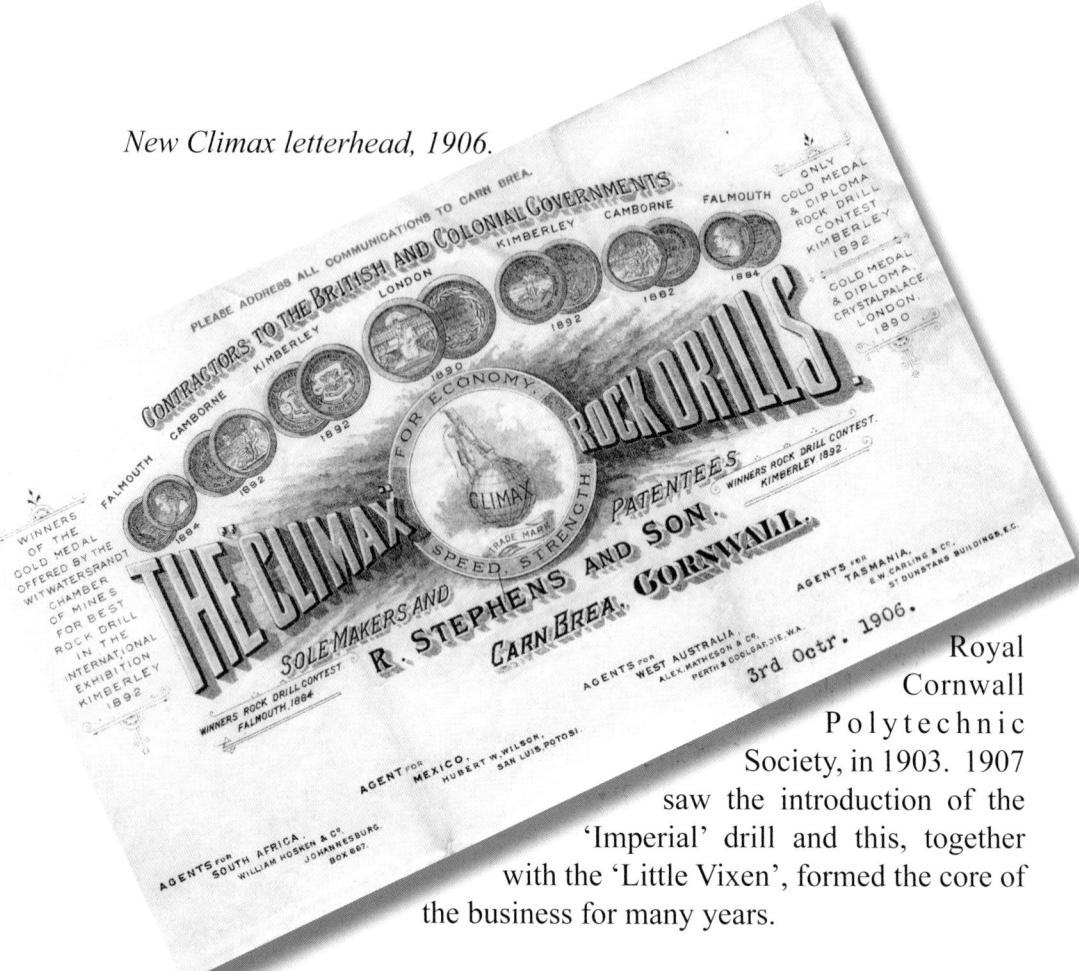

Royal Cornwall Polytechnic Society, in 1903. 1907 saw the introduction of the 'Imperial' drill and this, together with the 'Little Vixen', formed the core of the business for many years.

The Climax 'Imperial' hammer drill incorporated an improved anti-phthisis device. A report in the *Cornish Post* for November 7th, 1908 states that it was in use at South Crofty Mine, "being operated not by experts from R. Stephens & Son but employees of the mine, with excellent results." The drill diameter was 1¾ ins. and it weighed only 95 lbs (43 kg). It drilled a total of 93 ft in 4 and a quarter hours, a rate of 22 ft. per hour into hard Cornish granite.

On a narrow back stope on the 170 fathom level of Robinson's Shaft, with a load width of 2ft to 2ft 6ins, the drill bored 16 holes 3ft deep in just over 3 hours. With the softer ground experienced in the reefs of South Africa it was estimated that it would drill about 30–32 holes every 7 hours. Josiah Paull, the much respected manager of South Crofty, was "highly gratified" with the work of the 'Imperial' and said that "it would be absolutely impossible to work the stope with hand labour." The drill was operating in a space no more than 3ft wide and was cutting the lode clean out. By the 21st of November it was reported that in seven weeks, the cost of repair had amounted to just one shilling! This

as a fine spray onto the drill tip by means of a nozzle, using the compressed air that operates the drill to suck the water up from a suitable source, the volume of water used being adjustable. Clearly, the spray stops and starts with the operation of the drill, so no water is wasted and rate of water usage could be varied from one gallon of water every 5 to 80 minutes. It was therefore very economical in its use of water, which could be collected and recycled time and time again on site if necessary. The pressure at which the drill operated meant that water could be raised 25 ft. if required.

In 1906, William filed a patent (2141) for his cradle that supported the drills in position underground, to be followed by further patents in 1907 (8634 and 21586) and 1908 (7630 and 20404). This original cradle patent predates the Holman patents for cradles filed in 1907 (9243 and 9244) and was the subject of much dispute in the *Mining Journal*, William contending that the Holmans had patented a device already in the public domain.

William introduced his 'Little Vixen' drill, for which he won a medal from the

Climax rock drills being assembled, early 1900s.

Another Climax toolshop, early 1900s.

Following his father's death, William continued both to expand the business and be equally inventive, registering further patents in 1900 (7679 and 12458), 1902 (1244), 1903 (7979) and 1904 (9374 and 12626).

The 23rd of April 1904 saw the first appearance of patent no. 9374 for a device subsequently known as the 'Dust Allayer'. This patent was significant as it was the first design to allay dust as it was created by using water sprayed directly onto the drill tip. Previous dust allayers had simply been means of allaying dust in mines as a secondary operation. Entitled "Apparatus for Allaying Dust in connection with Rock Drilling Machinery", William's patent predates the patent filed by the Holman Brothers in 1906 (22191) despite their later claims to the contrary.

In rock drilling, the production of large quantities of dust is unavoidable. The first concern of the designer is to clear the holes as fast as they are drilled to avoid impeding further drilling. Prior to the Dust Allayer, air was used to clear the hole, forming large clouds of dust. These were inevitably inhaled by the operator to the detriment of his lungs and general health. The Dust Allayer introduced water

business enterprise had become in just 10 years and how their lives had changed. Richard Stephens died at *Havelock* on 9th January 1898 and was buried in the churchyard of All Saints Church, Tuckingmill on the 13th. William was away on business in Johannesburg at the time. Richard left his half share in the firm of R. Stephens & Son to his wife, Jane, together with £5,500. On her death, his four daughters, Janie, Nellie, Lucy and Blanche, would inherit. William, though he received nothing under the terms of the will, retained his half share of the business.

Patents and a developing business
Richard Stephens, and in particular his son William, were prolific in their designs and these gave rise to many applications for patents and trademarks. The first patent filed by R. Stephens & Son was no 4566 in 1883, subsequently followed by 2203 in 1885 and 2053 in 1890. These were followed in 1896 by nos. 1673 and 1674.

Interior of the works in the early 1900s, possibly part of the foundry.

included Dolcoath, East Pool, Wheal Agar, Wheal Basset, Cook's Kitchen, and South Penstruthal.

In 1890, Climax was awarded the Diploma and Gold Medal for their exhibit at the International Exhibition of Mining and Metallurgy held at the Crystal Palace, Sydenham.

A letterhead from 1892 shows the familiar Climax trademark and proudly proclaims the medals won in 1882 and 1884 from the Royal Cornwall Polytechnic Society as well as those from the Mining Institute of Cornwall from 1882, and the Gold Medal and Diploma of Honour gained at the International Mining Exhibition at the Crystal Palace in 1890.

In 1886 William had married Mary Anna Opie and a daughter, Dorothy Annie, was born in 1895 and a second daughter, Muriel Audrey, in 1902. A son, Richard Opie Stephens was born in 1886 but died in January 1888.

By 1891 both Richard and William had moved into houses in Roskear. Richard lived at *Havelock* whilst William and his family lived at *Endsleigh House* (later the site of the Camborne Registrar's office), an indication of how successful their

Panorama of the Climax works (and opposite page) from the early 1900s. The writing on photographs such as these was added afterwards, usually by someone who could spell!

The trademark 'Climax' was first registered in 1888. The names of drills were often very colourful and used to identify the various improvements as they were made. At about this time, above the gateway to the factory, an iron arch was erected, surmounted by a globe with the rock drill on it as in the trademark.

In advertisements in the *Mining Journal* of 1890, Climax drills claimed the highest awards in the last three competitions in Cornwall. In May 1890, one advertisement lists 10 gold mines, mainly in India, where the Climax drill had found favour and illustrates the international nature of the trade that Richard and William had generated.

Whilst South Wheal Crofty, the last and greatest of the local mines, had always been a major customer, another advertisement circa 1890, together with a list of testimonials, indicates the use of over 400 Climax drills in 12 overseas gold mines, (in addition to Mysore and Nine Reefs). Local mines using Climax drills previously made a 4-inch drill and so this required a completely new specification and drawings. William lost no time and although the enquiry had come at the start of the festive season, he worked through the night of Christmas Eve, all of Christmas Day and Boxing Day to complete both design and quotation.

Four days after receiving the enquiry, William, smartly attired in dress coat and top hat, paid his first visit to London, where he had an interview at the India Office with Sir Alexander Rundle, chief of Indian Railways. Sir Alexander inspected the designs and quotations and then inspected William, asking his age. "Twenty-five, sir," replied William. Sir Alexander reflected for a few moments and then said, laughingly, "You have had this enquiry for only four days, during which time Christmas has intervened, and yet you are here with the drawings and quotations. I like your cheek and your industry, and you shall have this order"! It is worthy of mention here that during the construction of the Kojak tunnel a World Record was set and commented on in the *American Mining Journal* at the time.

Interior of one of the toolshops in the early 1900s.

Climax crossroads around 1900. The GWR North Crofty Branch line can be seen crossing the road.

medals offered for rock drills, one for the drill itself and the other in the drilling competition.

The speed, strength and reliability of the 'Climax' drill soon became well renowned beyond Cornwall. The first foreign order was for 18 drills and was placed by Thomas Bewick, mining engineer for Nine Reefs Mine, situated in the Kolar Gold Fields in India. At about this time, Captain John Gilbert, manager of the Mysore Mine in India was sent to Cornwall to enquire into the merits of the various drills available at that time from all manufacturers. After several weeks deliberation he placed orders for their first drills with R. Stephens & Son.

In 1887, as a result of the excellent performance of their drills abroad, R. Stephens & Son were asked by Indian Railways to quote for a machine to be used in driving the tunnel to take the double-track Kojak railway through the Bolan Pass, in North West India. The tunnel stretched for a distance of 3500 ft. They had not

rather than one and renamed it the 'Hirnant'. So in total three machines, which were later used all over the world, emanated from Mr Richards' original design; a design for which he has received no recognition even to this day!

The rivalry between Richard and William Stephens with their 'Climax' drill, the Holman Brothers with the 'Cornish' and James McCulloch with the 'Rio Tinto' continued for many years and is best expressed in the many drilling competitions and challenges that took place between 1880 and 1900, and also in the correspondence, often acrimonious, in the *Mining Journal*. The rivalry was not all negative however, as it in fact served as a stimulus to both companies to strive for continuous improvement of their products. Richard Stephens used the motto "Just Drills", a reference to the fact that at that time Climax concentrated on perfecting drills whereas Holman Brothers had a host of other interests. Richard and William used Carn Marth Quarry extensively to test their drills.

Climax works at North Wheal Crofty around 1900. The road junction here became known as 'Climax crossroads'.

The first competition in which the 'Climax' drill was entered was held in 1882 at Camborne, but many more such competitions were held throughout the 1880s, culminating in the Exhibition at Crystal Palace, Sydenham in 1890. There, R. Stephens & Son won the Gold Diploma and Medal, and at the Kimberley Exhibition in South Africa, held in 1892, they again won both gold

Climax logo from 1890

Richard and William had visited a Mr Richards in a cottage at Wheal Agar and had seen both his design and a model for a rock drill but they concluded that their own design was superior, and so took no further interest in it. James McCulloch though did express an interest, so Richard and William took him to Mr Richards' cottage to see this model, which was exhibited with pride by the designer to James McCulloch.

Within a few days, McCulloch had joined James and John Holman in partnership to manufacture Mr Richards' machine, which they called the 'Cornish' rock drill. The partnership was short-lived however, lasting only two or three years. The split from Holmans was fractious as McCulloch had also contracted with Rio Tinto Mines in Spain during the partnership.

Following his acrimonious split from the Holman brothers, James McCulloch, visiting the Climax Works on a sultry August afternoon, came in to shelter from the rain. He related to Richard how the negotiations had been conducted, not face-to-face but instead through an intermediary, Holmans' solicitor, C. V Thomas, son of the legendary Captain Josiah Thomas, manager of Dolcoath Mine. Whilst John and James Holman were in his office, McCulloch was ensconced in the Commercial Hotel opposite.

McCulloch expressed regret at having to leave the drill with the Holmans, particularly as Richard felt that further improvements could be made to it. He pressed Richard, who suggested that the D-valve be replaced by a round cylindrical valve thus circumventing the joint 'Holman – McCulloch' patent. Soon afterwards, McCulloch introduced his 'new' drill called the 'Rio Tinto', in which he had incorporated Richard's suggested modification.

James McCulloch went into a new partnership with a company called Larmouth & Co., based in Manchester to manufacture the drill but, again, he soon fell out with them! Larmouth in turn modified the drill by including two exhaust valves

Address for Telegrams:
"STEPHENS, CARN BREA"

Awarded Gold Medal and Diploma of Honour International Mining Exhibition, London, 1890.
Medals Royal Cornwall Polytechnic Society, 1882-1884 Also Mining Institute of Cornwall, 1882-1888

R. Stephens & Son,

Wheal Crofty Iron Works,
Carn Brea,
Cornwall, March 11th 189:

"CLIMAX"

Engineers & Ironfounders

Sole Patentees and Makers
of THE "CLIMAX"
ROCK DRILLS,
As now supplied to
Her Majesty's Government
FOR
INDIA
Also to the leading Mines
Railway Contractors
at home and abroad.

COMPRESSORS,
CORNISH BOILERS,
RECEIVERS & AIR PIPES

Supplied on best Terms

OCTAGON OR ROUND "BORER"
DRILL STEEL and all kinds
OF IRON WORK.

head foreman), John Warren, Tom Letcher, George Green, John Penberthy, Alfred Tremelling and George Blight, all of whom served the company for between 40 and 55 years.

Many drills, manufactured by various companies, passed through the "Rock Drill Hospital." None of these drills was manufactured in Cornwall, however, and being designed for coal-mining, they could not cope with the hard Cornish granite. As a consequence they required constant repair. At that time there were several drills in use in Cornwall. Colonel Darlington manufactured a drill bearing his name as did Colonel Beaumont. James McCulloch had the 'Eclipse' whilst Mr Harris made the 'Champion'. Others included the 'McKean' and the 'Barrow' drills. Richard and William often incorporated their own improvements during repair and were astonished to find that new machines from the manufacturer would frequently feature these modifications. Richard approached one owner and offered to make rock drills for him. When he declined, Richard declared that he would manufacture drills himself, to his own designs. Thus the 'Climax' rock drill was born, with R. Stephens & Son becoming the first company in Cornwall to both design and manufacture rock drills. Naively, they did not secure a patent for this drill until 1883.

James McCulloch, a contractor using 'Eclipse' drills in Cornish mines, was a frequent visitor to the works. It so happened that it 'primed', drawing water and steam into the engine rather than just steam, thus destroying the power. Richard, being rather old fashioned, suggested cow dung and other possible remedies, but all to no purpose, and they were compelled, reluctantly, to scrap both boiler and engine. Another engine, previously used in some local streaming works at Treskillard, was acquired for £30 and installed. To their great satisfaction this answered their purpose perfectly, and was still in service in 1914.

William received no wages from his father, only occasional pocket money, but the other apprentices were paid sixpence a day. He was often sent to the various mines to undertake urgent repairs to drills, or to deliver spare parts, and went underground many times. On one occasion he was at South Crofty, arriving in time to go down with the seven o'clock shift. When he got underground the previous shift had not long finished blasting and the fumes from the dynamite were still lingering at the far end of the tunnel. In combination with an empty stomach, this caused William to be overcome by the fumes and he had to be dragged to safety by one of the miners, Andrew Harris.

A group of Climax employees in 1897.

The first employee of R. Stephens & Son was Joseph Blight, an apprentice, who later became engineer at South Crofty and East Pool mines. The workforce increased to include Harry Whitehead, who was employed for 53 years, (finally as

was an undivided moiety held jointly by Lord Robartes (later Lord Clifton) and Sir Hussy Freke.

Richard carried out many improvements so that the property served not only as the "Rock Drill Hospital" but also as a comfortable dwelling for his expanding family. They lived in the count house, which was commodious and comfortable, whilst the largest building was turned into the workshop. The buildings possessed neither doors nor windows so that they spent the next month, chiefly in snowstorms and gales, putting in doors and windows.

To equip this repair shop required the purchase of a steam engine to drive the necessary shafting and machine tools. They borrowed a horse with harness from Josiah Paull and walked to Helston, some twelve miles, where they purchased an engine from J. C. Toye, making the return journey on the same day with the horse dragging the engine and arriving home late at night.

To their great disappointment, efforts to refurbish the engine were to no avail. On lighting the boiler for the first time they discovered that

Awards to R. Stephens & Son in the 19th century. Mining Institute of Cornwall above and Kimberley Rock Drill Contest below.

would be *under* the cliffs, and accordingly he alternately scrambled over the uninviting rocks, or made wild dashes around headlands when the tide receded sufficiently for him to do so. Eventually growing tired of this, he decided to climb the cliff to reach Charlestown 'over the top.' Clinging desperately to every possible hold, he almost reached the top but then arrived at a point where the cliff overhung and was breaking away: he could go neither up nor down. William's plight was seen by a friendly coastguard who told him to go back at all costs. With great difficulty, he managed to clamber down under guidance, and once more waited for the tide to recede sufficiently to make another dash around the headland. He eventually arrived in Charlestown where he was admonished by his grandfather, Thomas, who told him that, to the best of his knowledge, only one man had ever before managed to reach Charlestown round the base of the cliffs.

Throughout his early employment, Richard harboured ambitions to run his own business. Recognising an opportunity to repair all makes of rock drills then being used in Cornish mines, none of which were then being manufactured in Cornwall, he borrowed £120 from his brother-in law, Samuel Crabb, and set up R. Stephens & Son. He leased the derelict count house of the North Crofty Mine, with several buildings attached and a yard of about one acre, all for a rental of £15 per annum. This property

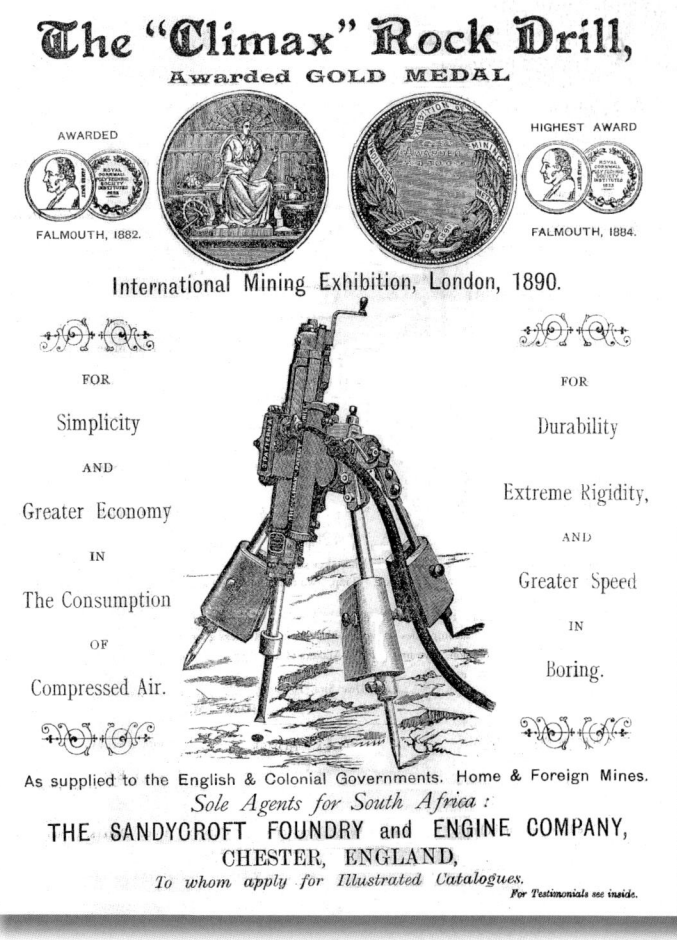

1890 advetisement for the new Climax rock drill

from St Austell, in December 1861 and in October 1862 William was born. The first of 13 children, only he and four sisters survived to adulthood.

Later, Richard moved his family to Pendarves Street in Camborne and was employed at the Carn Brea Railway Works, where locomotives and wagons were manufactured for the old Cornwall Railway Company. He later went on to work for Bartles, Tregonning & Duncan (later known as Bartles' Foundry) where he was in charge of the fitting and turning shops, and where William was apprenticed under his tuition.

Richard Stephens, a painting thought to date from 1890

As the eldest child, William was called on at times to perform the duties of nursemaid to his numerous siblings, many of whom died at an early age, and he often went to meet his father on his way home from work to acquaint him of the latest addition to, or subtraction from, the family. At school, William was neither clever nor industrious, being distinguished only for football, practical joking and a loathing of maths, the last of which he retained all his life.

His schoolmaster was a Mr Nettle, who later became a great friend and advisor. For many years Mr Nettle was head clerk at the Climax Works, until he emigrated to live with his married daughter in Australia where he died at a great age in 1921.

Holidays were often spent with his grandparents, Thomas and Mary Ann Stephens, at Porthpean. During one of these holidays, Charlestown was holding a regatta which everyone in Porthpean attended. William, though, had been left behind, so he thought that he would make his own way there. Unaware that he could reach Charlestown over the top of the cliffs, he decided the best route

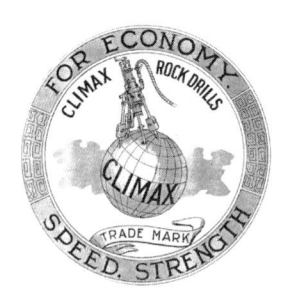

Chapter Eleven

R. Stephens & Son, the Rock Drill Hospital and the Climax Rock Drill

R. Stephens & Son was formed on 1st October 1878 by Richard Stephens and his son, William Charles Stephens, on William's 16th birthday. It traded under that name until 1913 when it became The Climax Rock Drill and Engineering Works Ltd. In 1951, it was acquired by Holman Brothers and run as an independent company but was gradually absorbed, becoming wholly owned in 1957 and finally closed as a separate entity in 1964.

Richard, the fourth of nine sons, was born in 1835 to Thomas Stephens, a coastguard in Porthpean, and his wife Mary. Seven of the nine sons survived and Thomas wisely apprenticed all of them to the Charlestown Foundry, each following a different trade. After serving their apprenticeship, several of the sons emigrated to America but three, Charles, Joseph and Richard, remained in Cornwall. Thomas Stephens was, incidentally, one of the Porthpean coastguards called out to protect the High Sheriff during the St Austell food riots of 1847. A man of many talents, Thomas, in his spare time, made and played many musical instruments. These included a flute (made from a fishing rod given him by the captain of a ship), to violins, cellos, and even pipe organs. He also made novelties such as snuff boxes with secret openings. Whilst his sons inherited his musical abilities they preferred more practical engineering, and Richard, after failing to persuade his father to set up a business at St Austell, worked at various small foundries in the neighbourhood.

An engine fitter, Richard spent his early years working as manager of the fettling department at Joseph Evans' Foundry at St Agnes. He had married Jane Crabb,

David Halfyard 1931-1996

A first-class cricketer, who included a spell as professional at Holmans in the 1970s as part of a remarkable career, was the late David Halfyard. Born in Winchmore Hill, Middlesex in 1931, Halfyard made his debut for Kent in 1956. A useful county player his career seemed tragically and prematurely over in 1962 when he suffered major injuries in a head-on car accident. He became an umpire but in 1968, after being seen bowling in the nets, he was engaged by Nottinghamshire. He played for them until 1970 when he joined Durham, then a Minor County, and later played for Northumberland.

1974 found Halfyard in Cornwall plying his trade with Holmans and he was very soon spotted and selected for Cornwall. He represented the county 27 times until 1982 and against Devon in 1974 he took 16 of the 19 Devon wickets to fall. He returned to the first-class umpires' panel while still turning out for Cornwall and continued to stand until 1996. David Halfyard died on 23rd August 1996. He was still bowling with some success for Tiverton Heathcoat CC a few weeks before he died.

Times cricket writer Alan Gibson found himself in Camborne in 1974 to report on Cornwall's match with Devon. "There was Halfyard at 43 still cutting them off on a damp Camborne pitch." Said Gibson: "He is one of the indestructibles, a kind of English Bill Alley, whom he resembles in physique, bowling action and humour."

Fittingly for one who led such a peripatetic career, Halfyard's pride and joy was his camper van with almost 400,000 miles on the clock. According to *Wisden Cricket Monthly*, "His career had the same improbable durability."

Holman bowls team in 1974, celebrating its silver jubilee.

Standing: R. D. Harris, W. Rashleigh, J. Walker, B. A. Richards, G. Thomas, R. Rickard, L. Truan.

Sitting: G. C. H. Lory, C. J. Selby (Chairman), J. Hancock (Captain), W. A. Dadow (Hon. Sec.), R. Merrit

Prizes: Dunstan Shield, Kenyon Cup, Registered 'A' Team Shield, M. D. L. Trophy.

In the 1960s the sport seems to have been very popular, taking up many column inches in the Holman Group Magazines. The building of the new administration office block in 1964 saw the loss of the bowling green at Dolcoath. The Sports Club Committee, with the help of the Company, purchased the Blaythorne site and included the construction of the present green with the other sports facilities. The Club was indebted to the Camborne Club which allowed Holmans to use its facilities for all league matches in 1994/96 when the current green was officially opened. A new pavilion was opened in 1996 and in that year a ladies section was formed.

the Rugby Football Union 1952/1953.

HOLMAN, N. P. T.: Director: Assist. Sec. Cornwall RFU

JACKSON, W. H.: Manager, Western Sales Area: Inter-county and International Referee, also played for Camborne

MATTHEWS, F.: Hollerith Dept (Cost Office): Camborne and Cornwall

MAY, L.: Foreman, P.T. Machine Shop: Camborne and Cornwall

PARNELL, Reg: Foreman, Grinding Dept, Climax: Camborne and Cornwall

RICHARDS, M.: Machinist, Maxam Dept: Camborne and Cornwall

ROBINS, G.: Packing Dept, Climax: Camborne and Cornwall

SELWOOD, Foster: Raw Material Stores, P.T. Works: Camborne and Cornwall

SELWOOD, C.: Grinding Dept, P.T.: Camborne and Cornwall

SELWOOD, Pat: Yard Foreman, No. 1 Works: Camborne, Cornwall, and England trial

THOMAS, George: Foreman Plumber, Carn Brea Area: Camborne and Cornwall

WAKEHAM, Gerald: Machinist, P.T. Works: Camborne and Cornwall

WAKEHAM, H.: Charge Hand, Fitting Dept, Climax: Camborne, Cornwall, and England trial.

WARMINGTON, D.: Fitter and Turner, No. I Works: Camborne and Cornwall

WARREN, T.: Goods Receiving Dept., No. I Works: Camborne and Cornwall

WILLIAMS, J. M.: London Director: Penzance, Cornwall and England

WILLIAMS, D.: Mechanical Handling: Camborne and Cornwall

WILTON, B.: Fitter and Turner, No. 1: Camborne and Cornwall

In 1960 the team was still struggling and after the opening of the new grounds at Blaythorne the rugby team played at Wheal Gerry until the beginning of 1966. During this period an appeal was made for more players; the team seems to have faded out during this decade, along with many other small teams.

Bowling Club

On 2nd May 1949 a bowling section of the Sports and Social Club decided to form a Bowling Section and a committee was duly elected. The first Chairman was Len Prideaux and Frank Hayman the Secretary. The club affiliated to the County in 1950. They relied on the generosity of the ICI (formerly Bickford-Smith) works who allowed them to use their green. With the formation of the Mining Division League in 1951, the need for a green of their own was considered as a matter of urgency. This was finally achieved in 1953 when a new green was officially opened on a site within the works at Dolcoath Avenue.

Climax personnel who had represented their town, Cornwall or England. This was an impressive list:

BIDDICK, W.: Inspector, No. 3 Works: Camborne and Cornwall
BUTLER, K.: Tool Room No. 3 Works: Camborne and Cornwall
CARTER, J.: Radial Drilling Dept, P.T. Works: Camborne and Cornwall
CARTER, R.: Shell Moulding Dept, No. 1 Works: Camborne and Cornwall
CARTER, W.: Grinding Dept, P.T. Works: Camborne and Cornwall
COLLINS, John: MAXAM Drawing Office: Camborne, Cornwall and England
COLLINS, Phil: Compressor R. & D., Climax: Camborne and Cornwall
FAVILL, H. L. V.: Director, Climax Rock Drills (India) Ltd: Redruth and Cornwall
HOLMAN, A. T.: Chairman: Camborne and Cornwall
HOLMAN, P. M.: Joint Managing Director: Camborne, Camborne School of Mines, and Cornwall. Hon. Sec. Cornwall RFU, 1924 1936: President of

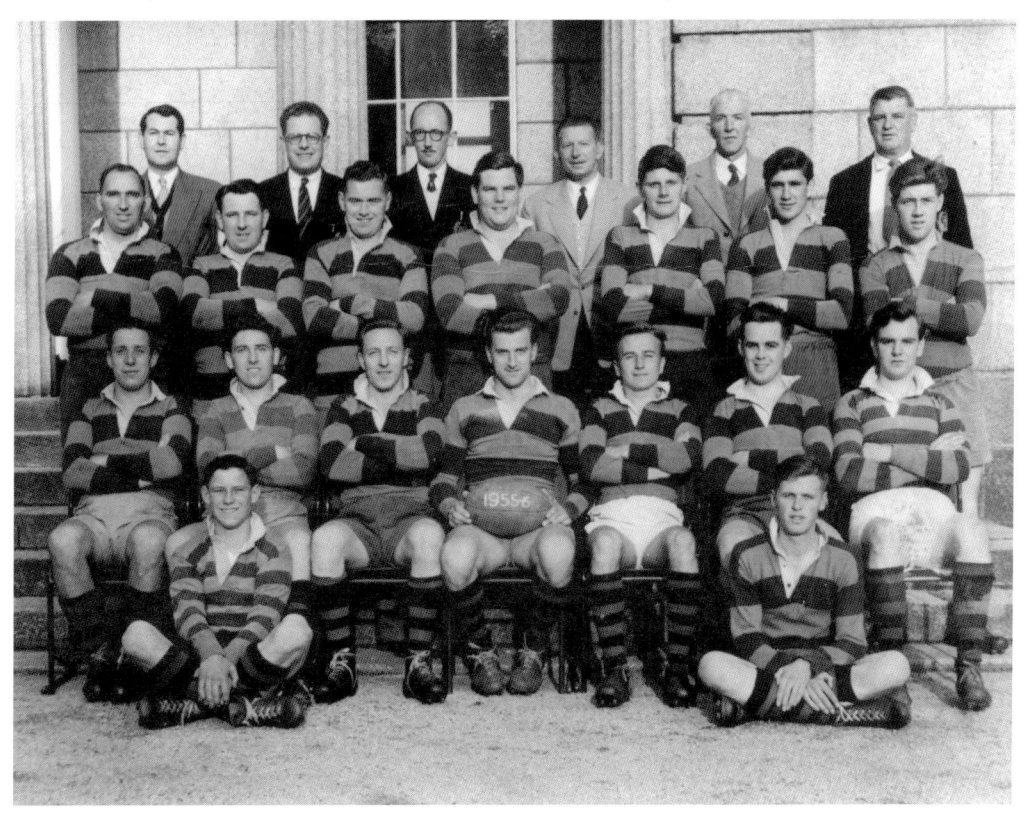

The Holman rugby team of 1956.

Rugby Club

Holman Brothers had a long history of rugby, and team players played in the original Camborne RFC after its foundation in 1878. Teams played during the early part of the 20th century, however there are no team reports in Holman Notes until the reorganisation of the Sports Club in 1948. Even so, in 1947 four Holman employees represented the County: F. Matthews, D. Williams, A. Solomon and K. Butler.

In 1955 the team lost only three games, one to Downing College, Cambridge; this was the first team to beat Holmans at home in two years. After this however the rugby team was affected by a lack of regular players; in the 1956-57 season the team put out a total of 44 different players! Members of the Troon Rugby Club helped out over several seasons. However there was a detrimental affect on the team which, in the 1957-58 season lost 12 games out of 17.

In 1958 the Holman Group Magazine published a list of names of Holman and

1950 rugby team.
BACK ROW: G. Clymo. G. Ford, J. Rule, H. Kemp, A. Wetherelt, J. Strike, D. Edwards, R. Trenowen, C. Selwood
SITTING: J. Laity, D. Cock, F. Tucker, C. Willis, C. Clemo, D. Teague, J. Williams
ON GRASS: B. Kemp, M. Pearce, J. Rollins

1956 Drop Stamping X1.

BACK ROW: *M. S. Thomas (Trainer), J. Ford, J. C. Honeychurch, C. Retallack, D. Passmore, W. A. Bennetts, C. Calloway, W. Beckerleg, A. Griible (Linesman)*

FRONT ROW: *T. Rule, T. Parr, R. T. Moon (Captain), D. Webber, W. A. Rule*

Chris Allen. This love affair with the Dunn cup continued. Following the first team moving to the Falmouth & Helston League, which they would dominate in the early part of the 21st century, the second team went on to win the Dunn Cup under the management of former 1st team manager Chris Allen in 2002, beating Illogan in a thrilling final The club then went on to retain the trophy in 2003 and 2004 under manager Mo Rogers. The 1st team now competing in the Falmouth & Helston League under managers Antony Reynolds and Andy Tasker would experience the most successful period in the clubs' history winning the league title and cup on three occasions and in their final season before successful application to the Jolly's Cornwall Combination League reached the final of the Junior Cup only to be beaten by the Eastern winners Morwenstow

Holmans returned to the Jolly's Combination league and senior football in 2004 and the 2005/06 season saw silverware return to the club once more in 2006 winning their 1st senior trophy the supplementary Cup. The club returned to the supplementary cup final in 2010 under managers Ray Richards and Neil Pitt losing narrowly to Falmouth Town.

until 1949, when the Holman Sports Club was founded, that Holman AFC was formed. In the late 1950s the team stuttered again, with insufficient players to keep the team playing regularly. The original home ground was at Wheal Gerry but when the new ground at Blaythorne was built, the team relocated there in 1964. The new pitch was used mostly for inter works games. In 1962 the sport was suspended owing to lack of players.

In 1966 the reformed team won the Hart Cup, under Chairman Leslie Vanstone, with a committee of Roger Johnson, John Symons and Alf Torbay. The club rejoined the Mining League and in their first season in Division 2 beat Carwell Green to lift the Hart Cup. The team had played 37, lost 10, drew 4 and won 23. The team seems to have been reformed again in 1968.

In 1983 Holmans entered senior football for the first time and spent 5 seasons in the Jolly's combination league until they were relegated in 1988. Having returned to the Mining League, the club began a period of sustained success starting in 1996 when the first team lifted the Mining League Dunn Cup under manager

1915-16 Munitions X1; what looks to be a professionally taken photo against a painted, probably studio, background.

and a second team entered the league.

The popular Andrew Cup competition was also a favourite; the team has won it on four occasions, 1979, 1988, 1999 and is current holder, though due to the competition rules it was unable to defend the title as it gained promotion to division two at the end of 2009 (only teams from division three and below could defend the cup). There have also been title wins in 1998 (division five), and 2002 (division four).

1st and 2nd cricket teams at Blaythorne, 13 June 1964

The second team had league title wins in 2002 (division seven) and 2009 (division six), making 2009 the most successful season for the club, as the first team also gained promotion to division two.

The team name changed again. After moving from the ground at Clijah Croft to Camborne School, and the club settled at Blaythorne, the home of Holman Social Club. After this the team lost players and eventually could not play. As Holman Brothers no longer had a cricket team it was decide to change the club's name to Crofty/Holman Cricket Club to keep the historic names alive.

In the 1960s a MAXAM team played regularly against both Works and local teams.

Football Club
Football teams have existed since at least World War I, when a Munitions XI was formed at Holmans, while during the 1950s a Cornwall Drop Stamping XI played. Teams seem to have been formed irregularly after that but it was not

Crofty/Holman Cricket Club. The team started playing league cricket in the Falmouth and District League in 1969. They then entered the Cornwall cricket league in 1970.

The club's most successful period followed, from 1980 to 1987, when they won the division two west title eight years running, only losing in two county finals against Bude and St. Newlyn East.

In 1987, following a break of the Cricket Club with the mine, the club name was changed to Crofty Cricket Club. Around the same time the club moved from its ground at Roskear (due to a housing development) to Clijah Croft in Redruth

The 1920 cricket team.

BACK ROW: C. Rowe, H. Alway, T. Eddy (Hon. Treasurer), W. Jewell, M. Trebilcock, H. Triggs.

SECOND ROW: C. T. Phillips, F. Pooley, H. G. Magor, L. Sullivan, H. Corey, B. Pemberthy, S. Merritt.

THIRD ROW: W. J. Lovelock, J. Moyle (Vice Captain), W. D. Corey (Captain), G. E. Timms (Hon. Sec.), F. Barnes, W. E. Butler (Hon. Gen. Sec.).

FRONT ROW: F. Rogers and T. P. Corey with the Cornwall Junior Cup.

Since the closure of CompAir, the club has had mixed fortunes. In recent years St Austell Brewery has made loans to maintain the club, but this cannot continue indefinitely. With insufficient income to cover costs debts have grown and this needs to be addressed. At the present time there is great interest in the Blaythorne site because it is the southern gateway to the proposed Treswithian urban development area. Sufficient land is available to sell to cover the club's debts and provide financial stability and to build a new clubhouse and bowling green on another part of the site.

Blaythorne House, shortly after completion in 1964.

Cricket Club

The first Holman cricket team was formed in 1920, however nothing is known about its early years. The Holman Sports Club was able to field a first and second XI for a number of years from just after World War 2. In 1947 five Holman employees represented the County: Wearne Cory, C. D. Rickard, R. Weekes, G. Holman and J. P. Angove. In 1957 two teams entered for Vinter Cup, also invited by Camborne Cricket Club to compete in Mining Division and the following year Climax entered two teams in the Holman Departmental Competition. Throughout the 1950s and 60s the teams were comparatively successful, but during the mid 1960s the cricket team merged with the South Crofty Cricket Club to form the

Section	Venue	Chairman	Hon Secretary	Hon Treasurer
Angling	Various	W. R. Stephens	P. S. Keverne	T. G. Cowling
Badminton	Canteen Hall, PT	K. T. Berryman	Miss G. Gilbert	Miss M. Sherlow
Bantam		L. Gundry	J. Kenny	L. Gundry
Bowls	Blaythorne	S. J. Dunstan	W. A. Daddow	W. J. Roberts
Male voice choir	PT staff canteen	E. Kessell	R. Dunstan	L. Campion
Cricket	Blaythorne	S. Woodward	J. Hocking C. Hocking	
Pigeon	Assembly Centre	R. Williams	B. May	D. Major
Rifle & pistol	Drill Hall, North Roskear	M. S. Moore	M. S. Moore	E. Harvey
Rugby	Wheal Gerry	D. L. Oates	P. J. O'Reilly	J. M. Cock
Soccer	Blaythorne	C. Haly	J. Jenkin A. Symons	O. Luke
Table tennis	Canteen Hall, PT		B. Angove	
Tennis	Blaythorne	K. T. Berryman	Miss E. Pryor	Miss S. Charles

Sections of the Sports Club, 1965

By 1965 club membership reached 2,000. Subscriptions were 2d per week for adults and 1d per week for juniors; salaried staff paid 10s. At this early date several sections had been disbanded through lack of interest. These included photography, philately, model aeroplanes and archery. A new section had been set up, for the company's anglers. The following year, finally, the bowling green was ready for use, having taken nearly two years to finish.

Two years later planning permission had to be obtained for an extension to the club house. Most of these sports seemed to have lasted for the life of the social club, being joined in later years by gliding and judo.

The Club Today

The clubhouse at Blaythorne is still a social club for former Holman and CompAir employees, though these decrease with time. The club is still home to football, cricket and bowls teams as well as social games such as darts and pool, though some darts players play in Taffy's Premier League.

sections of the club at this time included table tennis, badminton, bowls, tennis, cricket, rugby football, association football, modern dancing, photography, archery, bantam club and male choir.

To celebrate its opening, the site was visited by the sports clubs of Camborne ICI, as Bickford-Smith was then called, and the Climax Rock Drill Works. Bowls matches were played between the three companies and a tennis match was also played.

Percy pulling pints at the bar, his usual scowl almost forgotten in the excitement of the moment. Dick Holman pays for his beer while John looks on.

The official opening of Blaythorne's club house took place just before Christmas 1964, the ceremony carried out by Mr Percy. After making a brief speech to the packed audience he pulled the first pints for John and Dick Holman. John's toast was "to the Sports Club" after which he drank his beer from a tankard. During the evening Mr Percy recalled that when he played rugby as an apprentice in 1911 the field had a 13-foot drop across it! There had been fears that the opening was going to be delayed because of a legal hitch with the club licence application however the day had been saved by Leonard Smitheram who worked in the Carpenters Shop. Smitheram also kept the Commercial Hotel in Hayle and used his own Publican's Licence for the evening.

More funding was still needed to develop the club house and ground. The Tote Draw was successful though and the top prize had been increased twice to stand at a princely £6.

The building also required much work, fortunately well within the various fields of expertise required by Holman Brothers. Internal alterations were carried out by builders, and re-wiring by the various electricians. A number of less expert club members also turned out for miscellaneous do-it-yourself work, such as scraping, painting and plastering. Much of this work could have been called a 'Jan Luke' though it was actually official. To help fund the works it was decided to organise a Tote Draw and to collect 6d 'donations' each week, half of which was used to provide prize money.

'Jan Lukers' hard at work, adding the finishing touches to Blaythorne, late in 1964.

The new bowling green had six rinks, made of Cumberland sea-washed turf while the two tennis courts had hard surfaces. Bowling and Tennis Clubs had been formed to manage both sections, with financial assistance from the Sports Club. Only a few years old, nearly 1,000 of the company's 1,700-1,800 employees were members of the Sports Club. Membership cost 2d a week, allowing participation in sporting and recreational activities. The range of activities was extensive: Association football, cricket, bowls, tennis, badminton, table tennis, photography and modern dancing. Other activities later added were shooting, archery, pigeon fancying and athletics. A Holman/SWEB rifle club had been formed in about 1956 while the Climax Sports and Social Club still met separately, also having a 'Gay Nineties' club.

In 1958, following the take-over of Climax by Holman Brothers, the Climax Sports Club was formally incorporated into the Holman Sports Club. The various

Treve, always a keen sportsman himself, though Percy, as president of the Rugby Union, was away attending the final at Wembley. The site had been granted for a nominal rent and the work to complete the ground, including new tennis courts and bowling green, had cost several thousand pounds, borne by the company.

Percy's introduction to the new Queen at Twickenham in 1953 in his role as President of the Rugby Football Union.

Blaythorne House

The centralisation scheme that came about in the 1960s required the use of a large part of the Dolcoath Road sports grounds for the new admin building; this meant that the Sports Club had to move. Eventually a seven-acre site, including the former home of a mine captain, Blaythorne House, was acquired. This lay on the southwest side of Camborne, on the north side of the railway line. Work to convert the grounds to fields for several different sports commenced early in 1964. After a good deal of levelling, ploughing and digging took place and the new grass was sown in spring 1965.

Left: Holman Sports Club committee badge

Right: The enamel Holman Sports Club membership badge.

committees appointed. A new pavilion was built and paid for by the company.

Leonard Holman was elected President and the other directors are invited to be Vice-Presidents. The committee had a number of officers; these were:

Chairman: Wearne Cory
Vice-Chairman: N. G. Fiddes
Hon. Secretary: H. K. Milton (No. 1 Shipping Dept.)
Assistant Hon. Secretary: F. Tucker (No. 1 Cost Dept.)
Hon. Treasurer: W. D. Phillips (No. 3 Wages)

Representatives to the Executive :
G. Clymo, J. Hocking, A. H. Tregonning, R. Dower, J. Ford, E. Verran, A. Odger, J. Strike, C. Hodge, K. Berryman, H. D. Harris, D. Williams, H. Pascoe, L. May, F. Pooley, J. Floyd, P. Keverne and J. Roberts.

In response to a notice displayed on 28th February, 480 employees applied for membership within one month. By March 1949 plans for the ensuing cricket season were being laid, provision being made for continuing the popular "knock-out" competitions which been running for a couple of years. Secretaries of the rugby and soccer sections began preliminary arrangements for their next season; the table tennis committee was formulating plans for games in the Canteen, to be followed by badminton and possibly bowls.

The Holman 'Gay Nineties' Club (old time dancing) was also formed this year. Membership was by a yearly subscription of 5/-, with an admission fee of 1/3 per session; visitors were charged 2/6.

On 2 May 1953 the Club's new grounds at Dolcoath Road were opened by Mr

Demolition crew deep inside PCA, 12 December 2005.

Quaiffe Enginering: the final component completed, 8 December 2005.

The end of rock drill production in Camborne: the very last batch of rock drills.

Portable compresor assembly, 29th August 2003.

175

Holman Zoomtrack looking like some alien monster; unknown location.

Staff photo, 2001. Clive Carter at front left, with arms folded.

David Peckham and group posing next to Holman compressor at the Poznan Trade Fair, 1967.

View of the works looking north, probably taken in the early 1970s.

View of the No.1 Works Heavy Machine Shop in full operation. Looking into the works from gantry crane level men can be seen working at their benches while blueprints are seen on others. The shop is bathed in what is presumably summer sunlight through the northern lights in the roof.

Reproduced with the kind permission of the Cuneo Estate.

CONVERSATION PIECE.
This marvellous painting depicts the Holmans at the centre of operations at the works. The portraits show Nicholas, John, John Henry, James Minors and Leonard. On the left, Jack is in conversation with Paul, Jim is reading and John on the telephone. At the table Kenneth, Percy and Treve are discussing blueprints. Reproduced with the kind permission of the Cuneo Estate.

Rock drill trials in the Test Mine. Two drills are in use, a stoper, left and a rock drill with a stretcher bar on the right. In the right foreground a man is working on the compreessor.

Reproduced with the kind permission of the Cuneo Estate.

Nicholas Holman in deep discussion with Richard Trevithick over a Cornish boiler. While there is an engine house in the background this is artistic license as the scene appears to be in Holman's own works in Pool.

Reproduced with the kind permission of the Cuneo Estate.

The Sports Club badge

CHAPTER TEN

THE HOLMAN SPORTS AND SOCIAL CLUB

The early days

It was just after the Great War that a cricket club and tennis club were formed, while the rugby and association football teams had been in existence since the early 1900s. Despite this there was no attempt to regulate sports within Holman Brothers.

At the end of 1948 a deputation approached the Directors who readily encouraged the proposal to organise a Holman Sports Club. An 81 acre field near the Drop Forging Works (at Roskear) was placed at the Club's disposal and the Works Canteen was used for indoor activities. On the 18th January 1949 Percy Holman presided over the inaugural meeting when "The Holman Sports Club" was formally initiated and officers and

Guests of honour at the opening of the Dolcoath Road grounds, left-right: F. Hackwell, M. G. Bickford-Smith, Kenneth Holman, W. G. Bennett, Treve Holman and Dick Holman.

The Williams' Shaft winder model at Poldark in course of being covered with a timber building. The wire from the drum goes over a headframe outside.

the German machine first used at Dolcoath, the front of the Carn Brea Mine Holman Cornish Boiler, the Pendarves Estate clock and face, the old Redruth Town Clock mechanism and some models.

Following the final closure of CompAir in Camborne the Public Rooms have deteriorated rather badly and are in a poor condition. At the time of writing (Spring 2012) it has been accepted that the building would cost too much to conserve in its entirety however the plan is now to preserve the frontage. Work has now begun on the site to retain the Public Rooms facade to ensure this.

There are still some survivors of the museum to be seen. Probably the largest number comprises the Holman Rock Drill Collection, acquired by a mining engineer called Stephen Lay. This comprises:

Rock Drills - Holmans from 1881, and others, including a MacDermott and Glover (1860) and Doering (1867)
Early Models of Beam Engines etc.
Photographic Archives (glass slides and prints) of Holman products in use, for training and for demonstration
Brochures covering products and spare parts
Ledgers and various report books
Drawings including cutaways for sales, maintenance and training
Ephemera including various patents, diplomas, WWII war effort

The rock drills are currently on loan to the Trevithick Society and can be seen at King Edward Mine with a number of other rock and road drills which belong to the Trevithick Society.

The model of the winder for Williams' Shaft.

A model beam engine, made from steel, was removed from the museum prior to its closure by its owner over concerns for the leaky roof. This was returned to Cornwall in 2003 on permanent loan to the Trevithick Society and can be seen at Taylor's Shaft engine house at Pool.

A few items remain at Poldark, part of the original Holman Museum collection. These include the Holman model traversing winder, the bell from Ting Tang Mine, three rock drills including

agreement was confirmed in August 1980. However, in 1984 the Society's artefacts were taken to Geevor, following which some were brought back to Wendron, for reasons unknown.

Apparently Peter bought the title to the collection for £1 so that he had legal claim on the collection. It is not known why the price had dropped so precipitously to nothing; Peter Young already had his own site he did not need the building and its attendant problems.

Many records that came with the collection were lost in the late 1990s when left in a damaged building, a great tragedy. Although the roof had been badly damaged around 1996-97, repairs paid for by the insurers were not carried out and items in the room were not removed. The roof was not repaired until 2000 when the site was taken over by Richard Williams. Other documents had been removed by the official receiver and only a small number subsequently returned.

The Rostowrack engine being erected inside the museum.

Some items from the collection were apparently sold at auction while a number of items appear to have been abandoned around the site. This included a number of engines and sections of engines and early rock drills. These are sent to King Edward Mine as and when they are found. No list of Trevithick Society possessions deposited with the Holman Museum has been found but artefacts are known to be missing, for example a Clanny miners' lamp.

St Blazey foundry by William West. This engine is now in the possession of the Trevithick Society; currently it may be seen, dismantled, at King Edward Mine. However there are plans to erect the engine on site to be operated by compressed air.

A number of miscellaneous war-time items were also exhibited, from munitions to aircraft parts to the famous Holman Projector. The latter is now included with the Holman Rock Drill Collection, though it is not on display.

The question of closure of the Holman Museum was raised at the AGM of the Trevithick Society in September 1978. No more was reported on the subject until November 1979 when it was stated that CompAir had offered the building and contents to the Society for £40,000. At the time of writing it would be difficult, though not impossible, to raise such a sum; 43 years ago it could just not be done.

The Trevithick Society had had a long association with the museum via its long association with Treve Holman. Not only were there Society artefacts on show but the Society's council had met there for a number of years.

Close up of the rock drill display. In the foreground is a 3" x 5" bar hoist.

An emergency meeting of the Council of the Society had taken place following the AGM on 22nd September. CompAir had now sold its museum exhibits to Peter Young at Wendron Forge and was now asking the Society to remove its artefacts. Peter had offered to remove the Society exhibits to store at Wendron at no cost and hoped the Society would agree to their display there. This offer was accepted and an agreement would be drafted to that effect. The head of the Experimental Section, who suggested that these be refurbished to form the nucleus of a new museum. This idea was accepted by the management and it was Treve Holman who suggested that the old equipment be supplemented with new equipment to show entire ranges of Holman equipment. The Holman Museum was then set up in the Public Rooms.

The museum became a thing of great pride for Holman Brothers and visitors to the works, individuals and groups, were shown around. In addition, the various films made by Holmans were shown in the adjacent cinema. Many of these films were recovered from the CompAir site in 2004 (see page 330)

The museum held a large number of rock drills from the earliest to the latest Holman machines, other compressed air equipment a range of Maxam pneumatic control equipment and Goodyear Pumps equipment.

Many other items were on show. One important exhibit was the St Day clock, which had been used in Redruth for a century before being erected in a tower on United Mines, near St Day. The old timber carriage used by mine directors to travel on the Portreath tramway was also on show. This has since passed into the possession of the Trevithick Society and is on display at the Royal Cornwall Museum, in Truro. A number of models were also on display, including rock drills and steam engines; this included a model of the Levant man-engine.

Possibly the largest exhibit was the 22-inch rotative beam engine acquired from the Rostowrack China Clay Works. This was built at the

War time display: Holman Projector, depth charge thrower and Polsten gun.

Camborne Public Rooms

CHAPTER NINE

THE HOLMAN MUSEUM

The Camborne Public Rooms date from 1890, as shown on the foundation stone. In 1930 the rooms were acquired by Holman Brothers as a home for the ever-expanding Pneumatic Tool Works, better known as PT. As previously mentioned, PT occupied the site until the end of World War II, after which it moved to its final site, behind the No. 1 Works.

During the various departmental movements a number of items of old equipment, including old rock drills, came to light in odd corners of the various works. It was R. C. (Dick) Gilbert, then

The Museum in 1947. On show are: vertical steam engine which was originally used in the No.1 Works; display of rock drills including the Doering used in Tincroft Mine in 1867; sectioned Silver Bullet.

year, having retired in 1945. Starting at Holmans at the age of sixteen in 1886 he became Works Manager and was with the company for fifty-eight years. Mr Nicky had had a number of interests outside of Holmans, having taught mechanical engineering in Camborne; he had been a member of Camborne Urban District Council, Honorary Treasurer of Cornwall Rugby Football Union and a JP.

It was a considerable blow for the Holman family, yet they proved as resilient as ever and, despite the bleakness of the home economy, they still had their excellent foreign business. More new products were developed, like the 'Tractair', a small tractor mounted air compressor and, of course, the Silver Three, perhaps their most versatile and best known rock drill.

Kenneth Holman 1896-1954

Tuesday July 19th, 1955 saw a long service ceremony on the lawn at Rosewarne. Ninety-five men had been invited, thirty-six of whom were retired; 17 had completed fifty years and 78 between 40 and 50 years. This was a testament to the family nature of the company. For several years it had been the practice to give a gift of £50 to those who had

W. David Rule at his desk at the Welsh branch at Taffs Wells.

Rosewarne House, June 1951.

1951 Holmans bought the old Bartle's Foundry, which was largely used for maintenance work by South Crofty Mine and plans were laid for its development for lighter foundry work. The greatest prize of the whole year was revealed on 19 December 1951, when it was announced that Treve and Percy Holman had become directors of their once bitter rivals, Climax Rock Drill and Engineering Works. There was no longer the business to support such complementary firms in a Britain that faced 'gigantic economic problems' and Holmans were undoubtedly the more powerful of the two enterprises.

During the early 1950s the company began to modernise its office equipment; 'silent' typewriters had already been acquired and an electric one was in use by the Invoicing Department. A second electric typewriter was being contemplated by the Sales Department. Dictating machines had been installed at Rosewarne and it was possible to record telephone conversations onto discs. More advanced, a teleprinter had been installed in the Sales Department in Camborne and at the London and Sheffield offices.

In the midst of continuing success came the sudden loss of Kenneth Holman, who, in poor health for some years, died on 31 December 1954, having been at work at Rosewarne only the previous day. Another great loss to the company at this time was 'Mr Nicky', Nicholas Curry, who had died a little earlier that

Ode to the Silver Bullet
From Holman Notes No.15, June-September 1953.

And the miner's brow was sad,
And the miner's speech was shrill,
As darkly looked he at the face
And darkly at the drill.

"The boss will be upon us
Before one hole goes down
And then he'll truly slay us all —
With adjective and noun."

Then up spake one old-timer
(The daddy of the lot)
"There's but one thing to do boys
If you would fire that shot."

"Change to the drill I know of,
That's never let me down —
The Holman Silver Bullet —
Machine of great renown."

The miner went off scrounging,
Returning in high glee
He said, "I've been and found one
Pinched it from Number Three."

The hose was soon connected,
The steel was soon put in,
And soon on face of miner
Appeared a radiant grin I

The hole went down quite quickly,
Then followed several more
Whilst miner carolled gaily
And 'twixt the verses swore.

"Why didn't someone tell me
That work could pleasant be
If only you've a rock-drill
As good as the S.B."

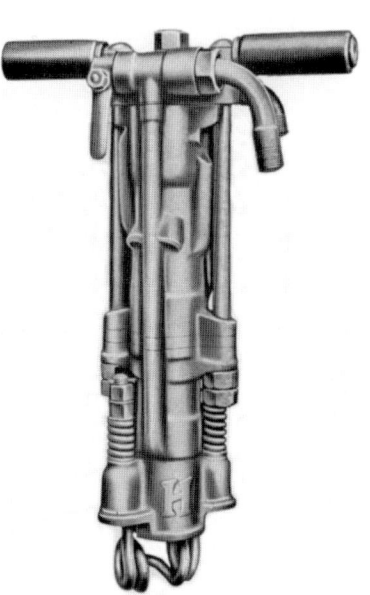

The Silver Bullet.

Industries Fair where they displayed rock drills and the celebrated cut away model of a Silver Bullet.

Beyond the celebrations it was business as usual and in June 1951 came the first of the radical changes for the company. The old office at Wesley Street had become far too cramped to cope with the rapidly expanding office staff and Mr Jim Holman, who did not wish to live at Rosewarne after his father's death, offered the house as a new headquarters for the company. Some had slight misgivings about the offices being so far from the works but Rosewarne was converted and, over two weekends, everything was shifted down the road, people happily going to work despite the tales of the headless horseman who supposedly rode up the drive on winter nights.

Lack of space was also a problem at No. 1 and in August

Staff enjoying the fourth celebration dinner on Friday 1st June 1951.

down in the canteen on four consecutive evenings, arranged according to length of service; something always jealously guarded by the old hands. Mr Treve, by now managing director, was invited by J. H. Reynolds to receive the presentation of a silver salver from William S. Rodda, while Alfred Angove handed Mr Percy a book containing the signatures of 1,870 employees. This book has survived and is in the safe-keeping of the Trevithick Society. William and Alfred were two of the oldest Holmans men but they could not rival Nicholas Curry, 'Mr Nicky', who was also present. Each employee also received a copy of *Cornish Engineers*, the anniversary book written by Bernard Hollowood, although Treve Holman himself collated and supplied most of the historical research into his company. Holman Week, which should have embraced a whole week of sports and entertainments including an appearance by the famous Dagenham Girl Pipers, was unfortunately washed out by the Cornish summer.

There were compensations as the 150th celebrations extended world wide to every corner of the Holman empire. Holmans won a gold medal at the Cornish

Examples of signatures in the Holman 'Red Book'.

others that the first ever 'Holman Week' should still go ahead in September 1949. Originally suggested as a way of creating money for charity, the Week also demonstrated the Holmans strong sense of responsibility for Camborne while emphasising the role of industry in a vigorous post war Britain. Surprisingly few townspeople had been inside the foundry gates, so highly popular works tours were organised, while the Holman canteen was filled every evening. There were parades, sports and a vast firework display. By happy coincidence, Holman Week also coincided with Camborne Shopping Week, both events giving the picture of a bustling industrial town.

John Leonard Holman 1887-1949.

1951 was a signal year for Holman Brothers Ltd. as not only did they celebrate their 150th anniversary but they also embarked upon an era of development, which slowly transformed the company into a thriving modern business. One success was the 'Dusductor' system of drilling, which was another product of the fertile mind of Treve Holman. Humidity in very deep mines prevented the use of water, so he came up with a system whereby the dust was extracted from the hole in much the same way as a vacuum cleaner and deposited well away from the working face.

Much of the 150th celebration reflected the 'fine family spirit' that had been one of their greatest strengths since the beginning. The directors had already announced that every employee would receive an extra week's pay so the employees responded by each donating a shilling towards the cost of a special presentation made to them by the directors. This presentation was made on 29 May 1951 in the form of four anniversary dinners when, in relays of 400, the work force sat

disembark and shove boulders under the wheels. Mr Treve, after an arduous trip to sort out technical problems with the Silver Bullets at the Aswan Dam, visited Holland and Germany before flying off to Canada to reorganise the agency. Mr Kenneth and his son Richard visited Lisbon and Barcelona while his other son James - 'Young Mr Jim' toured every mine, bar one, in the Gold Coast Colony and returned via the agencies in North Africa.

These extraordinary trips, which had been going on for years and would continue for many more, exacted a toll from among the now ageing Holman boys. Mr Leonard and his wife spent seven months travelling through Australia and New Zealand, which sapped his already frail health and he died on 1 June 1949. Lacking the fearsome energy of Treve or the great stamina of Percy, Mr Leonard was always a quiet and determined man, often self-deprecating, who despite a life spent in business and industry, found his greatest delights in the gardens and farm at Rosewarne.

Treve Holman became managing director of Holmans and he decided with the

Sleeve-lined Heavy Drifter, being checked at the Test Mine.

Rocks after failing to tow *Warspite*; and her sister tug *Tradesman* had 60 foot of wire wrapped around her propeller when trying to haul *Masterman* off the rocks. Aided by her compressor and two jet engines from an experimental aircraft the hulk was finally moved 130 feet closer to shore and by the summer of 1955 she disappeared from view.

A memorial stone was placed near the

Detail showing the rubbish-strewn deck of the Warspite and four of the Holman compressors.

sea wall at Marazion and later moved a short distance. The stone was unveiled by Admiral Sir Charles Madden and prayers were read by former crew member CPO S. J. ("Jeff") Barker, Honorary Secretary of HMS Warspite Association and a founder member of Winchester RNA. The remains of the masts lie in the yard at Porthenalls House, Prussia Cove and one portion was erected on a headland overlooking Prussia Cove. One of her 15-inch tompions is on display in the Royal Naval Museum in Portsmouth.

In March 1948 a drilling crew at Grootvlei Mine, South Africa, broke the World Underground Development Record using six S. L. 280 Holman drifters. In August that year the same crew broke the World Tunnelling Record when, at Marievale Mine, No. 7 Haulage, they made an advance of 1,277 feet in twenty working days using the same drills.

All the directors resumed their worldwide visits to the Holman agencies or far flung mining fields. Mr Percy had one especially adventurous trip when the troubled car he had hired to take him from Cordoba to a Spanish mine, finally burned out its brakes on a mountain road and all onboard had to hurriedly

Clyde. On the way, she encountered a severe storm and the hawser of the tug *Bustler* parted, whilst the other tug *Melinda III* slipped her tow. In storm force conditions *Warspite* dropped one of her huge anchors in Mount's Bay, which did not hold, and the storm drove her onto Mount Mopus Ledge near Cudden Point. Later refloating herself she went hard aground a few yards away in Prussia Cove. Her skeleton crew of seven was saved by the Penlee Lifeboat *W. & S.* For this outstanding rescue Coxswain Edwin Madron was awarded the Silver medal, and Mechanic Johnny Drew was awarded the Bronze medal. There were several attempts to refloat her but the hull was badly damaged and *Warspite* was partially scrapped where she lay

Warspite aground at Prussia Cove.

An attempt to refloat her in 1950, buoyed by 24 Holman compressors pumping air into her tanks, and watched by a large crowd, the press and the BBC, failed. There was insufficient water to float her clear of the reef in a rising south westerly gale. The salvage boat *Barnet*, standing guard overnight under the Warspite's bows was holed in the engine room, towed off and eventually drifted ashore at Long Rock, a few miles to the west. By August the battleship was finally beached off St Michael's Mount and after further salvage another attempt was made to refloat her in November. The Falmouth tug *Masterman* spent the night on the Hogus

Serving the World title frame.

Board. Holman equipment was helping build the Aswan Dam and the Owen Falls Power Scheme on Lake Victoria and all these new orders were invariably published in the new company magazine.

There were also the showcase exhibitions and trade fairs, which were already pointing the way towards a European Market and the Holman Stand was prominently displayed at Milan, Barcelona, Birmingham and many other venues. To ensure a wider audience, a company film was made in 1948. *Serving the World* was 45 minutes long and a comprehensive record of the work of every department. Lionel Marson of the BBC supplied the commentary. The film was reported in *Holman Notes*, a technical staff magazine which was the forerunner of the *Holman Group Magazine*, less technical, publications which ran for another fifty years or so. In 2010 the film was digitised and is now available as a DVD.

In July 1946, despite proposals to retain her as a museum ship, the Admiralty approved the scrapping of the Queen Elizabeth class battleship *Warspite*. On 19th April 1947, *Warspite* departed Portsmouth for scrapping at Faslane, on the River

Chapter Eight

150 Years Not Out

In 1945 Holman Brothers, like most of British industry, had to return to peacetime production and they appear to have adapted quicker than most, as the directors had rightly anticipated the post war boom in housing and civil engineering. There were inevitably difficulties, similar to those faced after World War I, with spares, material and manpower being in short supply while R&D had ceased in 1939.

The first and biggest task came late in 1945 when PT, the pneumatic tool shop, moved from the old Public Rooms into the vacated Polsten gun shop. Many of the original lathes and machine tools were taken over from the Ministry of Supply and one of the first automatic lathes was installed. A bonus was the conversion of a large part of the northern bay into the Holman Canteen, which catered not only for the expanding work force but became the centre for the social life of the company and Camborne itself. Everybody wanted to use what soon became known as the Holman Concert Hall. Six hundred could be seated and the small stage hosted everything from clowns, magicians and performing dogs to technical lectures and concerts by BBC 'Artistes' and, of course, Camborne Town Band whose director worked next door in PT.

Holmans did not require exhortations by the government to 'export or die' as, for half a century, three quarters of all they made had been foreign orders won against stiff competition. Production was soon up by 25% and during the single month of October 1949, a record 328 drills of all types were manufactured. There were fresh orders from India, Australia and Canada; 34 Silver Bullets for a Turkish Copper Mine, 26 TM 60 compressors for the South African Railways and 100 more Silver Bullets and another 26 compressors for the South Harbour Roads

of 'unassuming manner, imperturbable in temperament, as courteous as he was kind hearted, well read and well grounded in the study of national and empire affairs, he understood the good parts and other difficulties of socialistic and other political affairs.'

Sadly, he was just deprived of being at work during the royal visit when King George and Queen Elizabeth (later the Queen Mother) visited the works as part of the tour to Cornwall on 6th May 1942. They were met by Kenneth and Treve, who himself had been made High Sheriff of Cornwall, and it was noticed that the Queen wore on her lapel the coveted large 'H' brooch exclusive to Holmans. Holman Brothers' contribution to war production was amazing, considering that it was a relatively small plant in the far west of Cornwall. The Ministry of Supply alone had 1,350 air compressors and there were similar massive orders for rock drills, hoists, Rotopumps, air motors and pneumatic tools from the Admiralty and the Air Ministry. The Admiralty also had 475 depth charge throwers and 3,500 Holman projectors while the army received the 2,000 Polsten guns. Extra to all this was the production of oleo cylinders for the landing gear of Warwick and Wellington bombers and complete versions of the same for Spitfires. The machine shops also produced pins for Bailey bridges, Oerlikon gun casings, brackets for the caterpillar tracks of tanks, aircraft parts for Fairey Aviation and drop forging for the Fleet Air Arm.

Another loss to the company came in 1944 when the death of the London Manager, Arthur Herbert Smith, was announced in December. Smith had run the L. O. for 46 years and among other things had become Divisional Commandant of the City of London Police, Chairman of the Conservative Club at Stroud Green and a Freeman of the City of London.

King George and Queen Elizabeth walking along Foundry Road during their visit to the Works in 1942.

project, started in 1943, was cancelled in April the following year after it was realised that Allied air superiority over Europe made the vehicle superfluous; only three vehicles and eight conversion kits were completed.

Amid all this frantic activity came the death at seventy-three of Ben Cock on 1st April 1942, which ended another era as he really was the last of the rock drill pioneers who could recall the heady days of James McCulloch and the Cornish rock drill. Despite increasing ill health he had until a few weeks before remained at work as works manager at No. 1. His obituary summed up an amazing man

on 29th July 1943 the first completed Polsten was given firing trials at Porthtowan. Straight away there were more major problems as welded components failed under the shock of the fast 'blow back' system of the gun. More dangerously, the stop-double loading mechanism was faulty and had to be totally re-designed by Holmans.

By June 1944, a thousand Polsten guns had been built and within another six months no. 2,000 left W shop, complete, tested and with a spare barrel, while over 120,000 gun components had been turned out by the time the contract was cancelled that December. The Polsten was mainly intended to provide a highly portable and fast firing antiaircraft gun to support the Allied invasion of Normandy; indeed two of them were the only antiaircraft guns dropped with the airborne forces at Arnhem. Many were mounted on Liberty ships or used for airfield defence while the Canadian army evolved the 'Skink', a Sherman tank chassis boasting a quadruple antiaircraft mounting of Polsten cannons. The latter

N Factory complete and equipment installed; most of the workers seen here are women.

Early view of the construction of N Factory in October 1942.

office staff had to be re-directed to work on the new jigs, tools and gauges and then twenty major components had to be re-designed.

To accommodate the Polsten production line a brand new shop, the N or 'New Factory', with five bays and covering 65,000 square feet, was ordered by the Ministry of Supply. Holmans actually designed the new shop and supplied the builder with the gear to help erect it but there were considerable delays, although one bay was completed sufficiently quickly to serve as a lay apart store for the new equipment. Only eight machine tools were envisaged by the ministry but this soon rose to twenty-four, some of which had still not been wired up after six months, while hardening and inspection departments had to be added. An additional problem was the training of 'totally unskilled women' who had to be provided with mess rooms and general accommodation. To complete the wartime constructions a number of bomb shelters, to hold an official maximum of 275 persons, were dug around the periphery of N building. After a very fraught year,

Polsten gun being put on tow at a British barracks.

the water tower on No. 1 on a pitch-black night. One high spot of the platoon's history' almost led to a court martial but was fortunately preceded by a 'famous victory' when the garrison of Penventon was wiped out to a man.

The Camborne Works was vital in turning out components for tanks, depth charge throwers, Bailey bridges aircraft landing gear and trigger cases for Oerlikon guns and Bren gun parts. Personnel from the Royal Engineers and the RAF airfield construction units were trained on rock drills and air compressors both at the works and up at the Test Mine.

In 1942 Holman's was asked to manufacture components and produce main assembly drawings for the Polish-designed 'Polsten' 20 mm. antiaircraft gun. The original design work for the gun was smuggled out of Poland just before the German invasion and its production was placed with the British Stengun Company hence its name of 'Polsten'. It was intended to be a cheaper and far less complicated gun than the Swiss Oerlikon antiaircraft gun.

Holmans were already making parts for Bren guns and trigger casings for Oerlikon guns but they were unenthusiastic about the Polsten, correctly believing that the contract would interfere with their ordinary war work. Half the drawing

4 Aug 1941	HMS *Norland*	1 confirmed	Holman, 12 pdr. Hotchkiss & Lewis
July 1942	HML *141*	1 confirmed	0.5. Lewis and Holman
		1 damaged	
	HML *139*	1 probable	0.5. Lewis and Holman

and a strict blackout being enforced. Once again the benches were emptied as the Territorials and R.N.R. men were mobilised, while many men volunteered without waiting for conscription. The young apprentice who had signed up in 1932 finished his time on the Sunday and joined the Royal Navy the next day; on reaching Devonport barracks he was confronted by many of his mates from the foundry.

The works soon formed a Local Defence Volunteers Unit which was commanded for some time by Leonard Holman; the men parading in ordinary clothes and wearing armbands, with 'about one rifle to every twenty men'. Things improved when the Home Guard was formed, with the company being split into two platoons of 'B' Company 9th (Cornwall) Battalion. Sentry duty and exercises were readily done after a day or night shift and the Holman Home Guard was regarded as one of the best of the Cornish units. There was the odd incident when fear of heights led one sergeant, ordinarily an excellent soldier, to refuse to climb

Members of B Company 9th (Cornwall) Battalion Home Guard, disbandment photo.

Date	Ship	Result	Guns
1st June 1940	SS *Bombay*	1 probable	Holman & Hotchkiss gun
1st Aug. 1940	SS *Highlander*	2 confirmed	Holman & Lewis gun
8th Dec. 1940	SS *Treverbyn*	1 damaged	Holman & Hotchkiss gun
26th Jan 1941	HMS *Reids*	1 confirmed	Holman and Lewis
	HMT *Lord St. Vincent*	1 confirmed	Holman and Lewis
12th Feb. 1941	HMT *Eager*	1 confirmed	Holman & Lewis
27th Feb 1941	HMT *Lord Stonehaven*	1 damaged	Holman, Lewis & 0.5 in.
13th Mar 1941	SS *Instrahuir*	1 damaged	Holman, & machine guns
23rd Mar. 1941	SS *Beltoy*	1 probable	12 pdr. Lewis & Holman
6th Mar. 1941	SS *Andoni*	1 damaged	Holman, Lewis & Hotchkiss
18 Mar. 1941	FT *Thracian and Kingsway*	1 confirmed	Holman and Lewis
2 April 1941	HMS *Silver Dawn*	1 damaged	Holman, Lewis, 2 pdr.
9 April 1941	HMT *Avondale* & HMT *Lord Nuffield*	1 damaged	Holman, Hotchkiss 12 pdr and Lewis
27 April 1941	HMS *Patia*	1 confirmed	Holman, Hotchkiss, 3 ins. 2 pdr. & Harvey
27 April 1941	SS *Marie Dawn*	1 confirmed	Hotchkiss, Lewis & Holman
4 May 1941	MV *Araybank*	2 confirmed	Holman and Hotchkiss
18 May 1941	HMS *Corinthian*	1 damaged	Holman, Hotchkiss 12 pdr
	SS *Dalemoor* and others	1 probable 1 damaged	Holman, Lewis and Hotchkiss
28 July 1914	HMS *Typhoon*	1 confirmed	Holman, 12 pdr. Lewis and Hotchkiss

away, straight into the Oerlikons and Browning machine gun fire of Hitchens' other section which had circled around to lie in ambush.

Lieutenant-Commander Hichens, who won two DSOs and three DSCs, became an enthusiastic proponent of the Holman Projector and met Treve Holman on many occasions to discuss its development. On the Fairmile 'A' and 'B' boats the projector was fired by a control valve set on the bridge and upper deck and Hitchens had other refinements in mind. Early in 1943 while he and Treve Holman were dining in London, he sketched on the back of a menu card, designs for the Mk. 4 which would be less complicated and easier to train and fire. A prototype was built at Camborne and sent up to Dover but tragically it was never to be tested. On 16 April 1943 during a fierce fire fight in the early hours in the Channel, Robert Hitchens was killed by a single stray shot fired at very long range by the fleeing E-Boats. Hichens had been recommended for a VC for his actions in rescuing the crew of a Motor Gun Boat which was on fire. Hichens himself asked that it be dropped as he felt that he had endangered two boats during the attempt. The recommendation was re-submitted six weeks after Hichens' death, but rejected by the Admiralty for the reasons that Hichens gave.

For the rest of the war the Holman Projector continued to equip merchantmen and warships. Minesweepers, 'Flower' Class corvettes, tramp steamers and even tank landing craft disgorging Sherman tanks onto the Normandy beaches all carried projectors. There were light-hearted uses when bored trawler crews used their projectors to bombard each other with potatoes, hence its nickname of the 'Potato Gun'. Towards the end of the war the projector was further developed with the introduction of the 2¼" inch long-barrelled projector especially for flares. Large versions were built when the Navy found that the recoil from the throwers for their new 'Hedgehog' anti submarine bombs was damaging the decks of destroyers and corvettes. Holmans produced several, including one with a 7-inch bore which used a bigger cordite propellant and was capable of firing a 55 lb. bomb over 200 yards, though both prototypes were made redundant by the end of the war. One intriguing experiment was the mounting of a long-barrelled projector on a Rolls hydraulic recoil cradle in the nose of a Liberator patrol bomber. Once a surfaced U-boat had appeared on the bomber's radar, it could be illuminated at long range by a flare from the projector and a depth charge attack launched.

Back in Camborne it was very obvious by early 1940 that it was going to be a new kind of war with the works being camouflaged, fire watching being organised

to fire a 12 lb. rocket flare especially to take advantage of the weapon's near noiseless and flashless operation. Once at the desired height, the rocket ignited and flew for a mile and a half until the flare itself burst, burning at 600,000 candlepower for 45 seconds. Both the Mark 2s and the Mark 3s were modified for special tasks, some being given a 'bypass barrel' to vent excess air or cordite gas. This ensured a pre-set accuracy for the projectile, making it ideal for anti-submarine operations. Projectors were sent to Gibraltar to counter the successful raids by Italian frogmen and charioteers against shipping moored in the harbour. Incidentally they were not the only Holman equipment in action on the great fortress, as Holman SL9 rock drills, air compressors and drill sharpeners were then being used to drive galleries and defensive tunnels. Even in wartime Holmans achieved a record when an 8 foot by 8 foot heading was driven 180 feet in a remarkable 7 days.

Flotilla Commander Lieutenant-Commander Robert Peverill Hichens DSO* DSC**.

The Holman Projector became a truly offensive weapon during the battles between the Naval Coastal Forces and the German E-Boat flotillas in the close waters of the English Channel. It was a war of sudden ambushes and high-speed chases on black nights lit by flares or the quick flash of tracer fire. Not quite the theatre of operations for the sedentary projector but the weapons' potential was recognised by Lieutenant-Commander Robert Hichens R.N.V.R., a peacetime solicitor and international dinghy sailor who lived at Constantine and was also deputy town clerk of Helston. He saw that with only a light recoil and not requiring a heavy mounting, the projector was ideal for the lightly built motor gunboats then attacking the German convoys off the Dutch coast. These were always covered by the heavily armed E-boats and FLAK ships and Lieutenant-Commander Hichens, operating from Felixstowe under the Nore Command, evolved many tactics for thwarting the enemy. Indeed, he became the supreme tactician for the Coastal Forces. When going into the attack he divided his forces into two sections. One section used their projectors to fire flares above the E-Boats, which confused them as to their attackers' position, thanks to the projectors lack of flash or noise. The E-Boats invariably turned

with ammunition running low for both Lewis gun and 12-pounder, a gunner had desperately loaded the projector with three Mills bombs and caught the *Kondor* at close range.

The Holman Projector continued to be modified for a variety of roles on both land and sea. Being smooth bore it could fire almost anything from anti-tank rounds to the 'Molotov Cocktail', the celebrated petrol bomb in a glass bottle. Successful tests were also made firing a Schermuly lifesaving rocket line and projectors were adapted to fire grapnel lines to the tops of cliffs for the benefit of the commandos. Another development was the 'Chough' gun, a short-barrelled projector using cordite propellant and fitted with a gun shield. The Chough was capable of firing a wide variety of ordnance, including flares, grenades, spigot mortars and anti-tank rounds.

At the request of Captain 'D' at Devonport, the Holman Projectors were adapted

Wren N. S. Hopkins, from London, operating a Holman projector, on the Defensively Equipped Merchant Ship (DEMS) ranges at Cardiff.

The grenade case for the Projector. The top of the grenade can just be seen.

Holman was worried about the effect of the heat upon the Mills bomb. A test was arranged at Porthtowan when a steamroller borrowed from Cornwall County Council was connected up to a projector. The driver duly raised the pressure to 200 p.s.i, and the bomb was 'cooked' for twenty minutes without any ill effect and when it next fired a round, it performed perfectly, shooting the bomb over 1,000 feet high.

During the summer of 1940, Sir Winston Churchill himself demanded a demonstration but on arriving at the Aldershot range, Treve Holman and his team found the army had not provided any ammunition. The PM was already displaying his notorious impatience when someone recalled that there were bottles of beer in their picnic lunch hamper. The first bottle burst in the barrel but the next landed squarely on target, which provoked Sir Winston to jovially remark that it would 'save on cordite'.

From the time of introduction the Holman Projector slowly began to notch up a tally of German bombers, usually putting up a grenade barrage in support of fire from Hotchkiss or Lewis guns. A list of successful engagements survives in the Holman archive.

Treve Holman himself received news of one spectacular success for his projector.

> Dear Mr Holman,
> Commander Young has given me apiece of very good news to pass onto you and all working on the Projectors.
> A Holman Projector has shot down a 4-engined Focke-Wulf in mid-Atlantic. The machine, which was flying very low, was hit at the root of one of the wings and crashed into the water.....

The Focke-Wulf FW 200 *Kondor* was the long-range bomber which harassed allied shipping from bases in Brittany and was a deadly opponent, even for a fighter, let alone a ship armed merely with a primitive air mortar. Evidently,

Holman Projector also put out of action and William Birnie AB. and Bert Whyman, fireman, who were operating the same received facial injuries. The said injured men received first aid on board.

Resolutely, Captain Gifford resumed his voyage but at '…00.30 a.m., the vessel was again attacked by enemy aircraft, and two aerial torpedoes were seen in close vicinity which again passed close astern…gunfire was returned from the ship, until 00.50 am when aircraft was hit and crashed into the sea about 100 yards astern of ship'.

Battered and smoking the *Highlander* reached Leith at 10am with most of the wreckage of one plane still lying on her poop deck; so famous was the episode that Muirhead Bone, the eminent war artist was sent to sketch her. When news reached Treve Holman, he had the damaged projector sent back to Camborne where many years later it was proudly displayed in the Holman Museum. *Highlander's* Master, William Gifford was awarded the OBE, while one of the Lewis gunners, George Anderson, and one of the crewmen manning the Holman projector, Bert Whyman, were awarded the MBE. The incident was widely reported in the press and on the radio, so the Luftwaffe was determined to take revenge on the ship. *Highlander* was attacked again on the 13th and

Very poor image showing the wreckage of the Heinkel on the deck of the Highlander.

18th–19th September, but she escaped significant damage on both occasions. Later, *Highlander's* name was changed to *St. Catherine II*, when German aircraft attacked the ship again in the early evening of 14th November and her luck finally ran out; she was sunk with the loss of her skipper, thirteen crewmen and one passenger.

Within a month of the attack on the *Highlander* every trawler skipper was clamouring for a Holman Projector to reinforce his Hotchkiss or Lewis gun. Most of their vessels were steam and Admiral Leach rang Treve Holman asking if the projector could be converted to steam. This was feasible except that Treve

in which attack by Enemy Aircraft is possible' and to 'Keep air cock on receiver firmly closed until action begins. One bonus was that on exploding, the bombs left a heavy puff of black smoke which made the German bombers believe they faced heavy 'FLAK' and not merely a simple mortar made down in Camborne.

Coastal ships and patrol boats, even the RMS *Scillonian*, were fitted with projectors while naval crews and merchant sailors came from as far away as Stornoway to attend courses at Holmans, which included two or three days on the firing range at Porthtowan, where most of the test work was done.

A batch was shipped up to Aberdeen and within weeks came the first confirmed kill, during an attack on the North of Scotland, Orkney & Shetland Steam Navigation Company's steam coaster *Highlander*. Commanded by Captain William Gifford, she sailed from Aberdeen for Leith on the evening of 1 August 1940. She did not ship naval gunners but some of her crew had been trained at the D.E.M.S. school to operate her meagre anti-aircraft defences, a Holman projector on the foc's'le and an ancient Lewis gun in a sandbagged mounting on the stern. There was still a faint glimmer of summer dusk when at 11.45 p.m. the *Highlander* was attacked 6 miles south of Stone Haven by a pair of Heinkel He. 115 float planes, slow but sufficiently well armed to hunt the North Sea convoys. The second mate, seeing the white splash of a torpedo striking the sea, put the helm hard over. The torpedo passed close under the stern and Captain Gifford radioed 'A-A-A-A', signifying air attack, to Cullercoats Radio, which alerted the Royal Navy. The Heinkel banked around and came in again, her gunners raking the *Highlander's* port side. Her own gunners were ready and

Group of Mk II projectors on the cliffs near Porthtowan.

> Fire was opened by ship's machine gun and Holman Projector…at 00.10 am…Enemy aircraft which was still attacking and low over the vessel, was struck by gunfire from ship and crashed, partly on deck and partly in water in flames. Considerable damage was done to the ship's superstructure rails etc…lifeboats Nos. 4 and 5 were smashed, Wireless aerial put out of action,

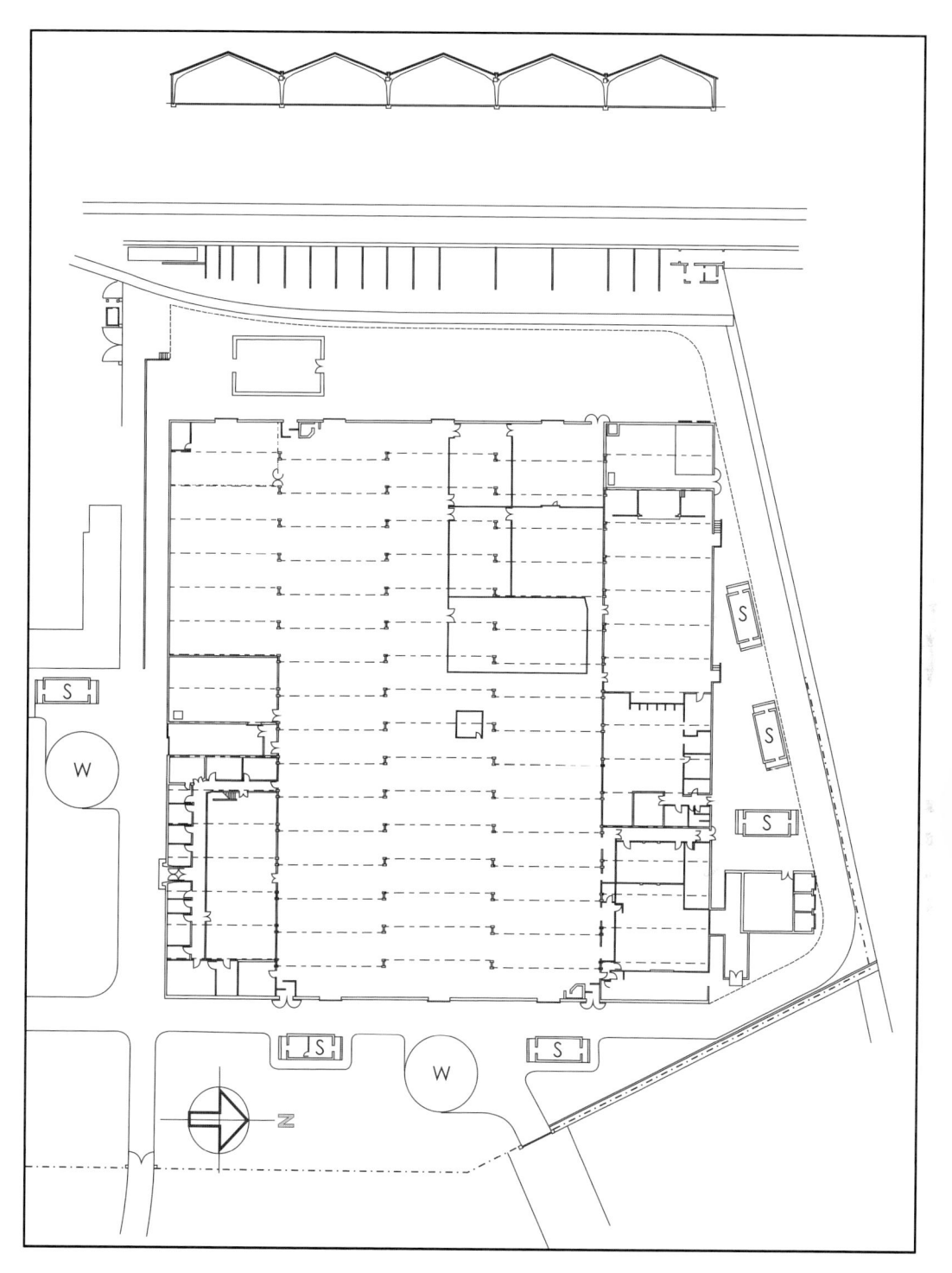

Plan of PT, 1941. Bomb shelters are marked 'S' and water tanks 'W'.

be seen. After a short time had elapsed and no explosion had occurred, a head appeared here and another there. Finally a rating crawled over to the bomb and dropped it into a bucket of water'. The Mills bomb was a dummy and therefore too light to trigger the firing mechanism but loaded with the proper ammunition the projector was highly effective.

There was an unforeseen sequel to the trial when Treve Holman called at the gunnery offices next morning to collect some prints he had forgotten. Stories of German spies were rife and even after being interrogated by naval officers for only ten minutes it still took two hours to get him back.

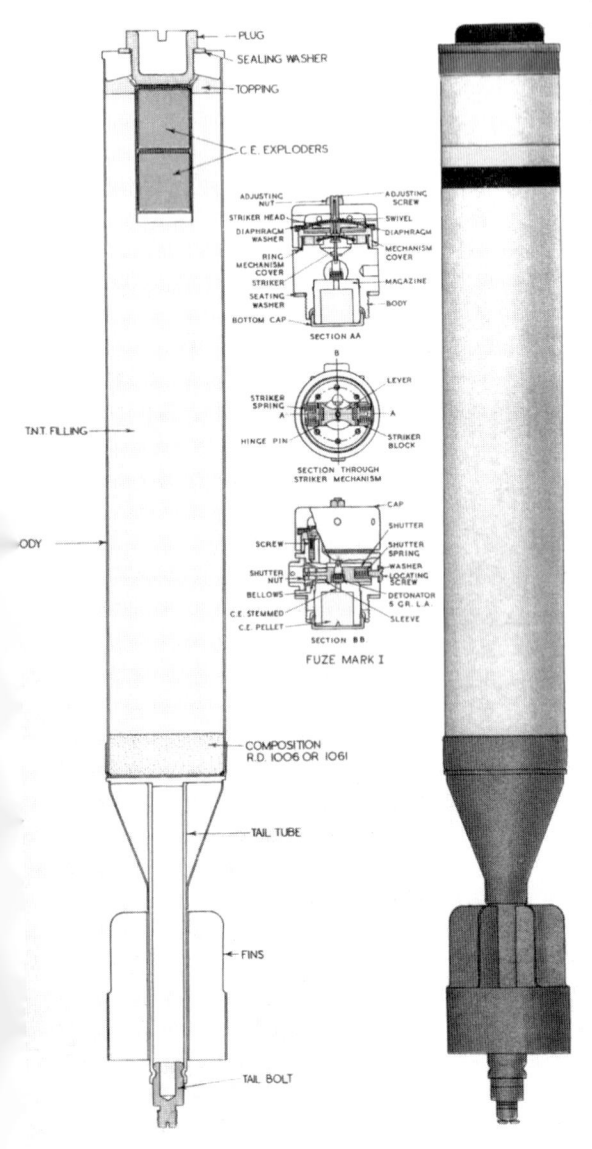

The trial led to an order for 1,000 but this was cut by half by the development of the improved Mark 2 Projector that was capable of firing two Mills bombs detonating at half-second intervals. Automatic firing was replaced by a simple twist handle, while the bomb canister was redesigned so that the lid fell away after half a second at about 100 feet, releasing the side lever and arming the missile. The straight bar sight was mounted at an angle of 20 degrees below the muzzle and for the first round it was pointed ahead of the target and the 'aim then corrected by observation of the flight of the bombs. Bombs should burst on or ahead of the target as bursts astern will have no moral or material effect. Gunners were warned to be careful of the controls and to conserve the air supply especially when 'in a locality

Diagram of the HE projectile for the Holman Projectile from the British Ordnance manual.

from material lying around No. 1. It was quite simple, with a steel tube welded to an air tank and fitted with a quick release valve, which was triggered by the missile. During tests a heavy weight was fired over twenty feet and, after further modifications, the mortar blew a heavier missile even higher.

Thus, by the autumn of 1939, the 'Holman Projector' had been born; one of the most unorthodox yet successful antiaircraft weapons of World War Two. Also, strikingly, it was not the creation of a group of 'boffins' or some sophisticated armament works but came from a relatively small works in the far west of Cornwall. Treve Holman still faced the daunting task of convincing the military, who naturally were always less than enthusiastic about innovative weapons. Fortunately the Holman Projector attracted the attention of Captain John Leach R.N., Director of Naval Ordinance, whose help and advice was to be crucial in the coming months. Commodore Young of the naval gunnery school on Whale Island at Portsmouth was sent to Camborne to inspect the projector and especially to discover its potential against low flying bombers. During the three days he spent at Holmans, the mounting was redesigned and rebuilt at his suggestions and after more tests another prototype was built. The specification demanded that the projector should be made only from readily available materials such as steel and cast iron and be capable of firing a standard mills bomb with a 3.5 second fuse to 1,000 feet, while the receiver had to hold sufficient for 50 firings at 30 RPM.

Testing projectors on the cliffs near Porthtowan. A young boy looks on.

In February 1940 the projector was tested at Whale Island where Treve Holman and his team found themselves ranged in a line with six other hopeful new weapons, each inventor avidly eyeing the competition. The first live firing tests were a fiasco. Having seen a rival's rocket round vanish towards Portsmouth, the Holmans men loaded and fired but there was only a faint click and hiss of air and the mills bomb dropped only 50 feet away. 'Within two seconds not a soul could

The Polsten 20mm anti-aircraft gun.

Chapter Seven

War Again: 'The Dreaded Holman Projector'

The war was only a fortnight old when the restless and inventive mind of Treve Holman evolved a way of converting compressed air into an offensive weapon. With his wide knowledge of technical history he knew the idea was not new; the ancient Greeks had first fired arrows with compressed air. The famous charge of Teddy Roosevelt's 'Rough Riders' at San Juan during the Spanish American war of 1898 was supported by a Simms Dudley compressed air cannon firing shells packed with Nobel blasting explosives. The U.S. Navy had tried a battery of compressed air guns on the USS *Vesuvius* while the Brazilians tested a large naval gun charged by a three-stage compressor. The first Whitehead torpedo was also propelled by compressed air.

Treve Holman, like his cousin Percy, easily recalled the Stokes trench mortar which was widely used in the trenches in Flanders, and he was convinced that compressed air was as good a propellant as cordite. In typical fashion, a prototype was hurriedly built

Switchboard operators at the Carn Brea exchange in 1940.

What is believed to be Holman's first order for China and the Szechuen Yunnan Railway in 1935; here are drill steel sharpeners and oil-fired boilers. Pictured in the Packing Shed with (left-right) E. Gundry, M. Trebilcock, B. Trevarthen, Leslie Blitchford, H. Pascoe and C. H. Phillips

extending into the Korean Peninsula. The province is now known to contain vast metalliferous deposits but in 1912 the first farms were set up to work the area's natural resources, mostly trees of various species. Initially Yunnan was connect to the port of Haiphong in Vietnam and not to other regions of China, and this was by means of a railway built by French engineers in 1910.

the first time came the realisation that such a uniquely historic engine should be saved for posterity. The cause was taken up by Cornish scholar Tregonning Hooper and Jack Trounson of Redruth, whose knowledge and enthusiasm for the Cornish engine was unrivalled at a time when 'industrial archaeology' was still forty years in the future. They approached Treve Holman, who was himself deeply concerned and well aware that Cornwall's industrial heritage was being swept away. With his enthusiastic help and support they formed the Cornish Engines Preservation Committee in 1935. The Levant engine was saved and the house made weather-tight while the public was admitted right up to the outbreak of war. These early pioneers of industrial archaeology could not have guessed that eighty years later the Levant engine would once more be steaming on the St Just cliffs and that the committee they began would still be in existence, now the Trevithick Society.

The year also included another high for the company: its first order for China, for the Szecheun (Sichuan)-Yunnan Railway; it included drill steel sharpeners and oil fired boilers. Yunnan is located on China's southern border, a lobe

Cornish Engines Preservation Society

ENGINES PRESERVATION FUND

LEVANT MINE 1823—1933

The original appeal by the CEPS for funding to help preserve the Levant engine. Members of the successor Trevithick Society later formed the 'Greasy Gang' to work on the engine.

The crowd at the unveiling of the memorial to Richard Trevithick outside Camborne public library in 1932. David Thomas collection.

1933. For over fifty years he had been 'A Camborne Captain of Industry' whose career embraced the advent of the rock drill and the development of Holman Brothers as a world wide company. His funeral cortege was headed by his sons and nephews and the Holmans men, while flags hung at half mast above the works which had been closed for the day. Yet his was not only the funeral in Camborne that day as, by tragic coincidence, Garstin Cox had died the evening after James. He had been unwell for some time but had come over from his home on the Lizard to nurse his father and caught influenza which became pneumonia. He was only forty one with much of his artistic career still before him, and his talent was recognised by Stanhope Forbes and by his old teacher Lamorna Birch. His funeral at Roskear Church was as large as James Holman, and attended by Ben Cock and his other uncles, C. V. Thomas and Mr Warren from Boiler Works while the bearers included Dick Gilbert and Percy Tromans.

It was about this time the Levant Mine finally closed down and those miners still left, began breaking the big collection of machinery at surface. The old Harvey pumping engine was speedily dismantled, though the bob fell inside the house and, becoming wedged fast, had to be blown up. A start was made on the Harvey winding engine that for ninety years had hoisted from Skip Shaft, but for virtually

thousands of which were in use in the Welsh collieries.

Although his health was failing James Miners Holman would live to see one his most cherished ambitions come to fruition. Always aware of his grandfather's friendship with Richard Trevithick and his family's roots in the industrial revolution he was determined to have a fitting memorial to the great engineer in his home town. In 1911 he with C. V. Thomas and J. J. Berringer, the principal of Camborne Mining School, established the Trevithick Memorial Committee. By 1914 they had raised £600 when it was decided to increase the height of the proposed bronze statue an extra foot to 7 foot 6 inches which brought the bill to £ 1,000. The outbreak of war had seen the quest laid aside but in 1917 a model statue was made and displayed at the Royal Academy, though there was no opportunity to make castings until peace came. The first memorial was a plaque commemorating the maiden trial of the Puffing Devil and set into the stone wall around Rosewarne at the bottom of Tehidy Road being unveiled during the Peace Day celebrations on 19 July 1919.

James Miners Holman had also proposed that the memorial should also include a 'Trevithick Scholarship' at the Camborne School of Mines and, during his visit to South Africa in 1922 raised £200 for the project. Eventually the whole memorial was 'completed mainly through the generous gifts and personal efforts of J. M. Holman'.

The statue of Richard Trevithick was officially unveiled in a memorable ceremony by the Duke of Kent on 17 May 1932. The Duke had landed in his aircraft at Tehidy and at lunch at Camborne School of Mines was introduced to the memorial committee including Leonard and Treve Holman, who represented their father. Early in the afternoon they escorted him around the works and then came the ceremony itself with a crowd of thousands surrounding Camborne Public Library where the bronze Richard Trevithick stood on a pedestal of Cornish granite. There was a guard of honour from the British Legion, and the Duke also inspected the Boy Scouts and Girl Guides to the cheering of 2,000 children lining his route. C. V. Thomas spoke at length on Trevithick and also the unstinting dedication of James Miners Holman in ensuring that the great Cornish engineer had a fitting memorial, then invited Leonard Holman, on behalf of his father to ask the Duke to unveil the statue.

James Miners Holman died at Rosewarne on the evening of Monday 28th February

V engine driving a colliery belt conveyor.

the advent of full-time staff in the 1950s it was largely serviced by staff from these companies.

Formed at the time of the Great Depression and continuing through World War Two and several recessions (including the current global economic situation), the Society has survived by continually evolving and responding to and meeting the needs of the membership.

Membership for the first 50 years was exclusively for UK registered manufacturers, but has in recent years expanded to include distributors, suppliers, users and individuals from around the world.

Holman Brothers, continued

By 1930 it had become 'imperative to find a new home' for the PT shops which were quickly transferred to the old Public Rooms adjacent to No. 3 and further expanded with new tooling, just as the Great Depression struck Camborne. PT barely managed to survive but production was eventually restored and, a new range of air tools and machinery were exhibited at the British Industries Fair in April 1932. The Soviet Union was by now becoming an important client and in 1931 the Russian trade representatives had put in an order for 11,462 drill sharpeners. Orders had also been placed with the Climax company down the road at Pool.

These included a special core drill, developed for the Ministry of Transport, to check whether the concrete used in making new roads had been perfectly laid. The 1936 Catalogue reflected the increasing importance of pneumatic tools, besides the innovation of 'Hot Mill' drill bit sharpening and the 'Vee' engines,

Mr Gaskell asked for more details about the scope and probable expenses of the society. The chairman replied that there was a somewhat similar society in America making such effective propaganda that it was necessary for British manufacturers to combat this, also the need for such a society had been impressed on us all by the Department of Overseas Trade, as well as the Lord Privy Seal.

At the inaugural meeting on 8th January 1930 it was unanimously agreed that E. C. Dunkerton of Robey & Co be Chairman. Following input from the Department of Overseas Trade, as well as the Lord Privy Seal, Mr. Mortimer of Peter Brotherhood proposed that the meeting should go ahead with the formation of the British Compressed Air Society. It was not proposed to have offices or paid officials as any business could quite well be run by the members. In this way the BCAS became a sort of precursor of CompAir; the twenty founder members of the Society included all of the future CompAir companies: Holman Brothers, Climax Rock Drill & Engineering, Broom & Wade Ltd and Reavell & Co. Ltd., and until

Crib time at No.1 Works in the 1930s.

Head Office staff posing outside the Head Office in 1935.

Institution of Civil Engineers. The first meeting was a continuation of the series of conferences of pneumatic drill and air compressor manufacturers held at the Department of Overseas Trade the previous year.

The main objectives of the Society were to promote and safeguard "by all lawful means the interests of the compressed air industry of Great Britain and Northern Ireland and to consider matters affecting it (apart from the relationship between employers and employees and the question of pricing arrangements)". Membership was open to all UK manufacturers of air and gas compressors, pneumatic tools, spray equipment, pneumatic control devices, and other compressed air apparatus.

Mr Dunkerton of Robey & Co Ltd was asked to take the chair. The definitive explanation of the raison d'être of the BCAS can be found in the minutes of the first meeting:

on the air valve drill and by the early 1920s an experimental pneumatic tool shop, forever to be known as 'PT', was established at No. 3.

There was a brief co-operation with F.T.A.C., an Italian compressed air company, but pneumatic tool development increased significantly when in August 1924 Holmans employed J. H. Arthur who already held several patents and had worked for Armstrong Whitworth and the British Pneumatic Tool Company at Coventry. Soon there was a production line turning out chippers, riveting and caulking hammers and the very successful coal pick which was used in thousands in the Welsh mines.

An Aside: The British Compressed Air Society
The British Compressed Air Society (BCAS) was formed in 1930 following pressure from Government on manufacturers of compressed air tools and other equipment. The primary purpose of the Society was to provide a forum for British manufacturers to consider imports and to encourage the manufacture of products for import substitution. This coincided with Government activity which led to the passing of The Merchandise Marks (Imported Goods) No 9 Order 1931. The inaugural meeting of BCAS was on Wednesday, January 8, 1930 at the

The SMS Hindenburg being salvaged in 1926. The salvage operations were interrupted by the outbreak of World War II and ships were again being broken up in 1946.

A Holman Road Ripper. great engineering battleships installed at Dolcoath and Levant. Above all they were mobile; just as Trevithick had put wheels on his 'puffer' engines and created road and rail locomotives, so Holmans had put wheels on their new powerful air compressors and opened a new era for civil engineering and themselves. New drills were needed and so arrived the Road Breaker and the Road Ripper which in the fists of large and lusty road men would often destroy the peace of leafy suburbia for the next seventy years. Road Rippers are still being manufactured today by the Holman-Climax Manufacturing Company in India.

By the 1930s the Type T double stage differential piston compressor had been introduced and, with various modifications and developments, remained the standard model until the 1960s. All this sophistication still relied on robust testing and the compressors were regularly towed at speed along the top of a cliff near Portreath by Holmans new Thorneycroft Lorry which had replaced horses in 1926. On a memorable occasion one of the spoked 'artillery' wheels came off, and breaking tow, the compressor slewed to a stop on the edge of Hells' Mouth. The new compressors found ready markets including the naval dockyards and were also used to help salvage sunken German war ships in Scapa Flow. In 1929 the Holman compressor won a Silver Medal and Diploma at the Royal Cornwall Polytechnic Society's exhibition, just fifty years after the Cornish rock drill was given its maiden trial by the same Cornish body.

It was at this time that Holmans embarked on the design and development of pneumatic tools and created another product that became synonymous with their name. Air tools had been in use for a long time and, like the rock drill, owed much to American inventive genius. By 1896 the superintendent of the Union Pacific Railroad's locomotive works at Omaha, claimed that in his machine shops 'air was as important as steam' while Naval dockyard and the G. W. R. workshops at Swindon already relied heavily on imported American air tools.

The development of the 'vane' type air motor made possible the application to an even wider variety of tools, and potential did not escape the Holman boys who had gained considerable experience of small machine and component manufacture through their war work. In 1919 Ben Cock had designed a concrete breaker based

air compressor flywheel being turned and a boring mill at work. Although the prince's tour was still being conducted at a 'smart pace' he paused in the yard next to Dolcoath Road to inspect two of Holmans increasingly popular portable air compressors, one driven by petrol and the other by diesel.

Half an hour later the Prince was on his way to No. 3 where, news having spread of the visit, so large a crowd had gathered that the police had to control the traffic. He was greeted at the main entrance by Ben Cock, the manager and his brother William and as no special arrangements had been made he met the men as if it was an ordinary working day. His interest was genuine and thorough as he progressed through each department until well over time he paused before leaving at the main entrance to watch Dick Gilbert bore 17 inches in hard granite in a minute with a stoping drill.

The portable compressors so proudly shown to the Prince of Wales belonged to a new generation of vertical high speed machines that had superseded the

The Prince of Wales and part of his entourage on their visit to the No.1 Works in 1926.

Engineering workshop in No.1 Works, 1928. This gives a good view of how the belt drives were worked by the overhead line-shafting.

Walter Peacock, Secretary of the Duchy, the prince jumped smartly out of his car at 3pm outside to No. 1 though the suddenness of his visit meant that there was only a small crowd. He was greeted by Treve, Percy, Leonard and Nicholas Curry and, after asking to be introduced to each foreman, was escorted into the moulding shop where he saw a chipping hammer and a patent moulding sand rammer at work.

In the pattern shop the Prince met the oldest employee, William Harris who had been with Holmans since 1901. He spoke with Sydney Nettle who had joined the company in 1895 and in the smiths shop met the venerable John Veale who, now seventy had begun as an apprentice with John Holman back in 1868 and whose own father had worked in the old foundry. He was not the only old hand as in No. 1 erection shop the prince shook hands with the foreman Harry Harvey, congratulating him on forty five years service before being shown an

reluctant apprentice, more so when his weekly wage rose to 4/- and then ½d was deducted for insurance. His father remained adamant that he should stay as 'a Holman indenture was worth its weight in gold all over the world'. The hours were long, 8.30 am to 5.30 p.m. with Saturday morning working. Half an hour was allowed for cleaning down the benches, which was often accompanied to the singing of men, who belonged to the Holman Male Voice Choir.

John Miners Holman, now nearing seventy remained outwardly the 'presiding genius' at Holmans and was still 'a gentleman largely esteemed by all who knew him at home or overseas and greatly loved by the staff he employs'. The grand old man of Cornish engineering received the ultimate accolade when the Prince of Wales visited the foundry on 20th May 1926. The Prince and the Duke of York had spent some time at the company's stand at Wembley, as they had two years previous, when he expressed a desire to visit the works but, his arrival was so sudden that Holmans were almost taken by surprise. Resplendent in plus fours and a fuchsia coloured tie and accompanied by Sir

A large compressor being put together in the Erecting Shop.

garden and planted 95 acres of woodland. Treve and Muriel Holman were keen gardeners who added many exotic plants brought back from plant hunting trips to the Far East and who created a woodland garden with advice from Sir Harold Hillier. Also present is a notable collection of magnolias, planted since 1945. Treve was very taken with magnolias and eventually had one named after him, *Magnolia campbellii* 'Treve Holman'. Treve also put his name to an annual cup awarded to the best magnolia exhibited in Cornwall. Additionally, the name 'Chyverton' also occurs in a number of Magnolia names as *Chyverton red* and *Chyverton white*.

A Lister 600cc petrol-engined Auto truck carrying crates between different parts of the Works.

When fourteen year old Harry Blackwell signed his indentures on 8 August 1932, they were no longer printed in elegant script but still signed by James Miners Holman himself. Young Blackwell already had a rich Cornish engineering background as his grandfather had installed Holman boilers in India, and his father, after serving an apprenticeship at Cox & Co. of Falmouth, had gone out to the Witwatersrand. There he was joined his brother, 'Uncle Charlie Blackwell', an ex-Harvey apprentice who, having relinquished his job at the London Waterworks, arrived at his parents door to inform them he was off to South Africa and, turning about caught the next train for London. He became chief engineer at Ferreira Deep and as a celebrated racing car enthusiast, won the M.C. by leading a motorised column into German South West Africa in 1914.

Young Harry whose later career was more adventurous than his uncles', was a

disabled servicemen and were ready to employ more to fill their government quota. This was typical of the Holman boys who saw their civic responsibilities as inseparable from their own business. To promote the skills and prospects of their apprentices they were given the opportunity to sit the exams which brought prize money, diplomas and free evening classes at the Camborne Mining School. One of the earliest scholars was George Lawry, the son of Sergeant Major Lawry, who had joined No. 1 stores in 1912 and after service in France, returned to operate the head office telephone exchange. His other two other sons, both apprentices, were already working in the U.S.A.

It was this year that Treve and his wife Muriel moved into Chyverton House at Perranzabuloe. The house was built in 1730; two wings were added in 1770 by John Thomas, a wealthy mine owner, who also created a Georgian landscaped

Machine shop in No.3 Works, 1924.

One of the early Holman portable compressors, this one in 1924.

became known as the 'Test Mine' and in a notebook, written in faded pencil there such entries as 'Jan 1926 Tested new 50 lb. Light Drifter with vent washer. Will not stand any strain all air blowing through vent holes' or '24 Jan. Light Drifter came back from Cumberland tendency to stop when pressure put on it. Valve has not enough air to blow it over when back'. The writer was Richard 'Dick' Gilbert who at thirteen had been apprenticed to Harvey of Hayle and when that foundry closed had with another young man demolished a 135 foot chimney. He joined Holmans as a fitter in 1904 at No. 1 but the next year emigrated to the U.S.A. and on returning in 1907 went to the Rock Drill shop at No. 3. One of his earliest jobs was to go up to Millom in Cumberland and install new condenser work and pipes on a 70-inch engine but he soon displayed an absolute genius for rock drill development and he and Mr Treve were largely responsible for opening the Test Mine.

By January 1924 the *Mining World* reported that the 'famous mine engineering works - of Messrs Holman at Camborne, there is work in abundance even to the extent of entailing overtime employment in some departments'. There were now 532 employees and, unlike many Cornish firms, Holmans had taken on fifteen

Holman Brothers paypacket from 1926. The halfpenny has never been taken out, though it is now wearing a hole in the side.

servicing Nigeria, Sierra Leone, the Cameroons and the Gold Coast.

A new development was the 'S. L.' or 'Stream Lined' drills when, to eliminate projections like the valve chest, the valves were placed at the rear of the machine, ensuring quicker action for less air. New circular valves replaced the tappets and steel balls and 'kicker ports' were introduced to speed up the valve action and increase drilling speed. The air winches were also developed into newer models, the largest capable of hoisting a ton vertically, or double that up a 30 degree slope. The stretcher bar hoists had come far since the days of James Holman's shipboard chat with the homeward bound mining engineers; the most powerful could raise a ton and a half, or drag 3 tons up a very steep incline. One innovation was the Heal-Shaw Beecham air motor which was capable or working even under water, and was fitted to a range of Holman equipment including the 'Superhaul' where it was mounted inside the winding drum. Much of the new equipment was displayed at the British Empire Exhibition held at the Wembley Exhibition in May 1924. The Prince of Wales and the Duke and Duchess of York spent considerable time at the impressive Holmans exhibition stand and the company subsequently produced a magnificent souvenir brochure which was the first ever pictorial record of the whole works.

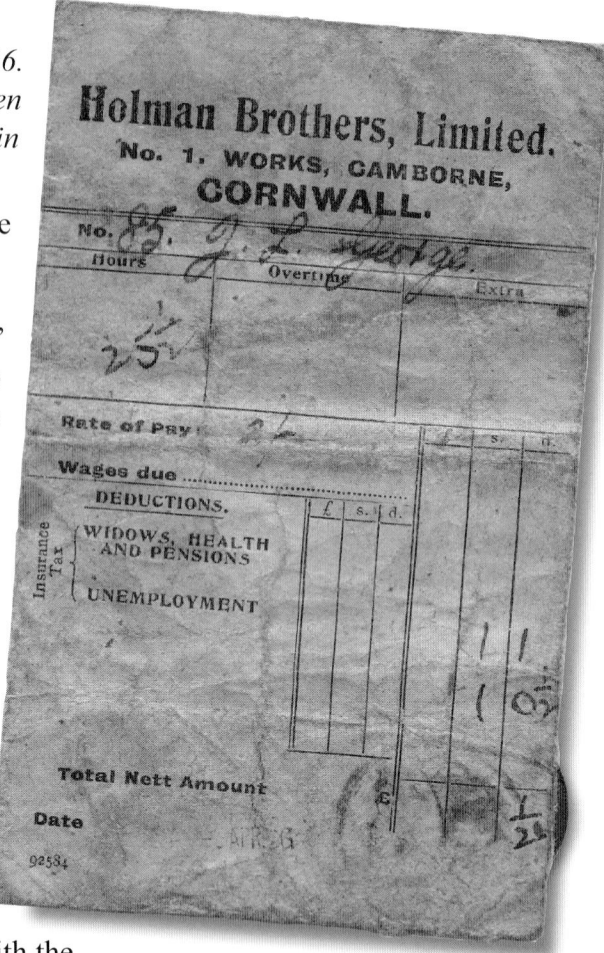

Holman guaranteed that every piece of equipment was thoroughly tested before leaving the works and to ensure this in 1920 they acquired a small granite quarry near Blackrock at Troon. Hundreds of rock drills were tried out in what always

American companies now monopolised the gold mines though Holman 3½" drills had established another record by sinking the main shaft at Crown Deep in July 1919; four drifters and two jackhammers sank the South African equivalent of Willams' Shaft 279 feet in 31 days and then followed up with a stoping record at Van Ryn Deep. The Holman agency was still managed by C. F. E. Vivian, who, after mining in East Pool and on the Rand, had joined them in 1911. There were then only three employees, supervised by J. H. Vivian and the agency operated as a department of Fraser & Chalmers Ltd.

Following an interview with Victor Williams, overseen by Mr Nicky Curry, lucky young men were allowed to become Holman apprentices. For the first year these lowest of the low earned a princely two shillings per forty-seven hour week. However, there were deductions, one shilling and a halfpenny per week for National Insurance and, every third week, an extra shilling for the Holman Group scheme. The four weeks pay looked like this:

Week 1	2s	1s 0½d	11½d
Week 2	2s	1s 0½d	11½d
Week 3	2s	2s 0½d	-½d
Week 4	2s	1s 0½d	11½d

Total for four weeks: 34½d (2s 10½d)

In fact it was illegal to pay a negative amount, so the apprentice was advanced a halfpenny in week 3, that is, pay was zero, and this was subtracted from week 4!

In 1921 the company opened an office in Garth House, Taff's Wells, Cardiff, to deal with enquires from the Welsh coal fields, the first of Holmans provincial branches. The young W. David Rule was given charge, and was still Branch Manager over twenty years later.

Mr Percy, after a trip to India, came out to the Rand early in 1922 and spent a year in Johannesburg organising the business. By 1927 Holman South Africa (Proprietary) Ltd had been established and developed into a large engineering works capable of supplying machinery and spare parts. Mr Leonard went off to Canada and returned home in time to leave for Spain with Mr Kenneth who had been made a director in August 1922. The important West African goldfields were not neglected and by 1933 a branch had been established at Takoradi where workshops, staff bungalows and Cornwall House had been built to house staff

Atmospheric photo of a small portable compressor being used in Trafalgar Square by the Acme Paving Company in the 1920s.

nephews were convinced that the future of Holman Brothers lay in developing new products and customers and the pursuit of foreign orders was to remain the almost sacred duty of the directors for generations to come. More immediately the task was to regain those customers lost through the concentration upon war work and the Holman boys resumed their frenetic sales trips all over the world. Treve Holman had barely returned home from flying Sopwith Camels over Flanders, when, after pausing to get married and abandoning the opportunity to stand for Parliament, he sailed for South Africa, together with Ben Cock, just as W. C. Stephens returned after a successful trip to the Witwatersrand. Treve was to make frequent visits to the gold mines and once returned with a painful souvenir as, when laughing over some joke in a stope, he was struck in the teeth by a flying pebble.

modern mine, even 'if the Cornish engine persists in its reputation as evidenced by the attitudes of the management of South Crofty and East Pool'. The electric pumps were raised from Williams' Shaft, which was stripped of its machinery but the amazing Holman winder was too unique to be re-used.

The steam winder from Harriet Shaft, though elderly, was used, being rebuilt and given an overwind trip and speeded up to hoist at 2,000 feet per minute. Holmans engineers dismantled the cross compound air compressor at Williams' Shaft and later the big lattice head gear which was tailored to fit Roskear Shaft as a replacement for that from Harriet that was used as temporary sinking head frame.

James Holman's unremitting faith in Cornish mining was never really rewarded as both the Tolgus tunnel and New Dolcoath were expensive failures. His sons and

Aerial view of No.1 Works from the 1920s. The presence of a tram in the background indicates a date before 1927. It is interesting to see the lack of development south of the Roskear branch line.

James Holman's steadfast dedication to Cornish mining was demonstrated only a few months later when, undismayed by the loss of the Tolgus Tunnel, the East Pool committee decided to sink a shaft right where it should have ended. The Tolgus Shaft Company was formed and in May 1922 Holman drills began sinking a 17 feet by 8 feet shaft among the old Tolgus Mines. By December the shaft was down 500 feet and was to sink another 1,500 before cross cutting out to the south-west but it was one of the biggest and most expensive failures in Cornish mining. At Dolcoath the drive under the Roskears had not been abandoned, even when the mine finally shut down in April 1921. A new company was formed with Holman and his son, Leonard, among the directors and early in 1923 R. Arthur Thomas announced that New Dolcoath would be sunk in the heart of the Roskears. Piggotts the Welsh pit sinkers, who had put down Williams Shaft a decade before contracted to sink the new shaft, which was slightly smaller than the latter though also brick-lined.

The remains of the old Harriet Shaft winder in its house at Roskear. The drums are mismatched and one is presumed to be a replacement.

By July 1924 the new shaft was down to 400 feet and R. Arthur Thomas was determined that he would not have some redundant eighty or ninety erected on his

items of the engine viz. the bob and the cylinder are in position' though it was October before the Holman pitwork was in position. There were other delays until on 20 November 1922 Captain Paull recorded that 'We only ran the engine about half an hour yesterday afternoon to see if everything was all right and I am pleased to say both engine and pitwork was as near perfect as one could wish, no leakage of steam anywhere and in the shaft the pitwork was very steady'.

Early work on the Roskear Shaft in September 1923.

Like thousands of other companies, Holmans used large numbers of horses, from the big shires for hauling heavy equipment to the cab mares that drove the family members around. Outside working hours these were allowed to exercise in the fields around the stables. The horses were controlled by carters such as John Brown, 'Farmer' Vivian, Jackie Bawden, Nicky Saundry and Alf Simons. These men would think nothing of harnessing up a team to take a huge cast-iron beam across the country to its destination.

To feed these animals the local fields were seeded with hay, wheat and corn. At harvest time the 'Thrasher' arrived it was a case of all hands on deck to get the cereals threshed in time, the hot and dusty work being aided by copious amounts of dandelion and burdock as well as 'herby' beer.

The great beam of the 90-inch engine being hauled by traction engine from Fortescue's Shaft. Clive Carter's father is apparently part of the crowd.

The catastrophic collapse of East Pool did provide some extra work as, threatened by serious flooding, Captain Paull of South Crofty promptly wired the liquidator of Wheal Grenville not to sell the 90-inch pumping engine on Fortescue's Shaft. Privately he reckoned 'it would be worth while to secure the plant if only on spec as the sum I have valued it at is practically breaking up price'. Holmans were now the only foundry capable of making the pitwork and heavy castings required for a Cornish pumping engine. On 18th March 1922 Captain Paull wrote '... I told Holman Bros, yesterday to get on with the pitwork anyway. They hope to deliver the set for the 195 in 4 or 5 weeks and the other sets as we want them at 1 or 2 week intervals'.

Captain Paull's bargain buys, including two steel Lancashire boilers, whose price he had managed to 'whittle down' arrived behind a pair of traction engines in May 1922. The installation of the 90-inch engine and the new engine house of shuttered concrete were supervised by Joe Blight, a former Climax apprentice who had worked on the installation of the Severn Tunnel pumps but it was a laborious task. By early August Captain Paull was able to report that the 'biggest

Holman hammer drills.

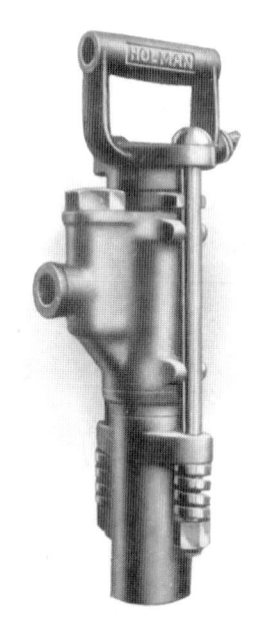

downwards. Shafts were blocked, including North Whim which was worked by the Holman beam winder, and worse all pumping ceased. The flooding threatened South Crofty, although down in the Tolgus Tunnel, 1,600 feet under Illogan Churchyard, the Holman crews faced a flow of only 25 gallons per minute.

East Pool appeared doomed to become just another 'knackt bal' but James Holman and his directors reacted with a swiftness rare in Cornish mining. To regain Roger's Lode they would shift the mine to the opposite of the main road and on 21 October 1921 they commenced sinking the new Taylor's Shaft, which in 20 months was due to bottom out at 250 fathoms. East Pool used their own miners and the new Holman HD3 hammer drills, which, to the delight of Captain Taylor, easily put down 20 foot holes in the hardest greenstone and would have done double that if required, a response that soon appeared in Holmans publicity.

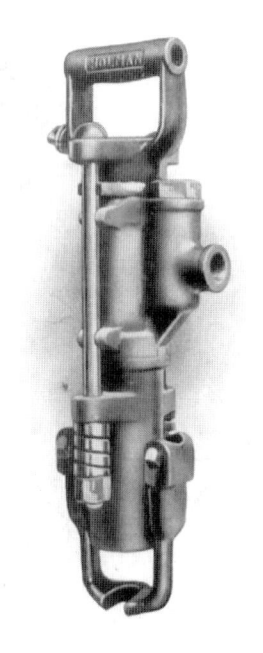

The fortunes of East Pool contrasted sharply with the rest of the 'Black Country' around Camborne for, even as the Holman drills bored steadily downwards, the depression had worsened. Carn Brea Mine had finally closed in January 1921 and Captain Josiah Paull had reluctantly put South Crofty into care and maintenance Things were so bad that the Cornish Electric Power Company deprived of many good customers, was uncertain about supplying the surviving mines or even the Holman foundry and the tram cars running between Camborne and Redruth. The Holmans men themselves willingly donated to the relief fund for the unemployed miners while the Holman Male Voice Choir would soon embark on a tour to raise money for the same fund. Later John Holman himself had to lay off men from throughout the works until orders revived.

returned to normal, though work now started at 8am after the half an hour for breakfast was abolished. A cricket club and tennis club were formed, the rugby team beat St Ives while Ben Cock's artistic nephew, Garstin Cox, having returned from war work had a painting accepted by the Royal Academy and was soon being commissioned by Holmans for more calendars. In reality Holmans quickly discovered that sacrificing normal business for war had a bitter dividend with most of their overseas markets hijacked by the American mining and rock drill companies. The company's only lifeline was new products and hard work by both employees and directors but the same did not apply to the Cornish mines that were once their good customers. Basset Mines collapsed on Christmas Eve 1918, though Major Frank Oats would not see his great mine dismantled as he had died at his daughter's house in Port Elizabeth in Natal a few months before. Only the Holman air compressor and the winder survived, the latter going up to a Somerset coalfield where a Holmans man saw it at work many years later.

Even the great Dolcoath was tottering though James Miners Holman, like R. Arthur Thomas was deeply reluctant to see the 'Queen of the Cornish Mines' go under. There was little they could achieve, except await the outcome of a deputation to try and secure aid from a government that had more enthusiasm for profitable coal than uneconomic tin. James Holman was still chairman of East Pool whose wartime profits from tin and wolfram enabled the mine to embark on the kind of development denied to Dolcoath. It was believed that the very rich Roger's Lode, discovered in 1915, ran eastwards under the old Tolgus Mines, which had never been sunk deep enough to strike the tin. Holman and his fellow directors soon formed the Tolgus Tunnel Company and in February 1919 four Holman 3¼" inch drifters, mounted side by side in a battery, commenced driving a 10 foot by 8 foot heading 2,000 feet northwards from the 255 in Old Wheal Agar Engine Shaft. The urgency of the work was reminiscent of Brakpan or Van Ryn Deep; miners worked continuous six hour shifts and to save time 'mucking out after an end was blasted, the broken rock was allowed to fall into a steel net which was hauled clear by a Holman air winch and dumped into empty stopes. A double tram road was installed and a smiths shop and rock drill repair shop cut out of the solid rock.

Driving through increasingly hard granite, the tunnel at 860 feet cut through a very rich lode of tin and wolfram then came disaster. Early in May 1922 there were small roof falls in the Great Gunnis at the 180 in East Pool; these increased and became a huge subterranean landslide as the old copper workings collapsed

The Johannesburg factory circa 1950.

Chapter Six

Recovery and Revival

'Tourists and others passing through the town of Camborne by rail or road usually express astonishment at discovering a large engineering works in this part of the country. Their astonishment is considerably increased when they stop to investigate and make a tour the various shops of the four sections of the works...'

Within weeks of the Armistice the lucrative war contracts had been precipitately cancelled without any opportunity for companies to readjust to peace time work. Holmans speedily informed the *Mining Journal* that they were 'now in a position to receive orders for general machinery'. Life at the foundry slowly

Holman Brothers letterhead from 1918, probably new for that year and the changes it would bring.

Ipatiev House, the site of the murder of the Imperial Russian family in 1918. In 1974 it was formally listed as a Historical-Revolutionary Monument. But, to the embarrassment of the government, it was steadily becoming a place of pilgrimage for those who wished to honour the memory of the imperial family.

In 1977, as the sixtieth anniversary of the Russian Revolution approached, the Politburo decided to take action, declaring that the house was not of 'sufficient historical significance', and ordering its demolition. The task was passed to Boris Yeltsin, Chair of the local party, who had the house demolished in July 1977. Despite this, the pilgrims kept coming, often in secret and at night, leaving tokens of remembrance on the vacant site. After the fall of the Soviet state the Church on the Blood was built on the site, now a major place of pilgrimage.

jewellery and the walls pockmarked by bullets, later being told that the bodies had been flung down a derelict mine shaft.

Arthur Thomas left Yekaterinburg as the Bolsheviks again advanced and on arriving home, related his experiences to Herbert Thomas, the editor of the Cornishman. Herbert, never able to resist a 'scoop', printed the story of the Czar's execution. This was reprinted in the *Pall Mall Gazette*, mainly because the editor was a Cornishman but it was ignored by the national newspapers which did not carry the story until several months later.

Howitzer bombs	Track links for caterpillar tracks
Lancashire boilers	Trawler engines
Mine sinker parts	Tripods for Maxim gun mountings
Newton gun beds	Wombat boring machines

Things were getting more difficult with shortages of food, materials and manpower, as men were being conscripted. Supplying machinery to the Cornish mines was near impossible, let alone to the new mining companies which were ransacking every long forgotten seam for the vital wolfram. Captain Josiah Paull, who was endeavouring to restart Castle-an-Dinas Mine, desperately needed a sand table, which required a priority certificate to enable Holmans to interrupt their production line. They did 'promise to proceed with the construction of the table as soon as they have the order, even without a certificate but much prefer to have the latter in case of a munitions inspector calling and possibly stopping the work for something which he would consider more important'.

War production continued apace with special orders for the 'Wombat' hand-powered drilling machine and the Holman tunnelling pump, both of which were used during tunnel warfare on the Western Front. The Wombat had been used by Australian tunnelling companies at Hill 60 near Ypres, having been invented by Captain Stanley Hunter of the Australian Tunnelling Corps. They were hand worked and relatively silent so as not to reveal the position of the tunnelling companies, many manned by Cornish miners, who burrowed deep under the German lines to lay enormous charges of high explosive prior to an offensive.

Many men, like Jack Holman, wrote home about their experiences during the war, but few could rival the extraordinary tale of Arthur Thomas, a first cousin of the Thomases of Dolcoath. After managing the El Oro Mine in Mexico and a copper mine in Argentina, he went off to Russia as a representative for Holmans, who had a large shareholding in a mining company in Siberia. By July 1918 he was in Yekaterinburg which was full of Bolshevik partisans and, as Honorary British Consul, he happily stored the valuables of fellow expatriates and many of the White Russians, risking being shot if this was discovered. The deposed Czar Nicholas and his family lived out of sight behind a tall palisade in a small villa, but as the 'Czech Legion' and the White Russian army advanced, Arthur Thomas and everybody else realised that something sinister had happened to the Czar and his family. He was among those invited to inspect the basement of the villa where he was horrified to see blood splashes, fragments of burned clothing and

celebrated Holman Minstrel Troupe and the Holman Quartet that regularly appeared at benefit concerts for good causes or supporting the war effort. Many girls found that, though long and arduous, war work was an exciting and liberating alternative to the drudgery and poor pay of domestic service or farm work.

The girls and young women excelled as machinists on lathes, shapers and grinders and readily acquired the exacting skills demanded by having to inspect the assembly of intricate components like gauges or gun sights. The more robust even tested the new light jackhammer drills before these were sent to South Africa. All kinds of girls went into munitions, including the mother of Clive Carter. She was barely seventeen when she found herself assembling shell fuses in the big wooden-hutted works set up on Camborne cricket ground by James Holman's old friend Captain Bennetts of the Roskear Fuse Works. A tribute to the Holman girls came in the company's calendar for 1917. This showed a 'charming young girl in khaki trying out a part of a Holman rock drill, while the Union Jack forms an effective and patriotic background.' 'Carry On' is the title in the footnote. 'Not withstanding the war, production of Holman rock drills is steadily carried on.' Neither was the birthday of James Miners Holman forgotten when he received a presentation signed by his staff and the heads of departments.

During the Great War, Holman Brothers manufactured a vast number of items for the war effort:

Bickford Patent cap presses	Nitric acid retorts, eggs and mixing tanks
Bomb gauges and jigs	Paraffin engines
Broaching machines	Powder feed boring heads
Cordite forming presses	Powder moulds with needles
Cordite mixing machines	Proof shot
Cordite presses	Punches and limit gauges
Cornish boilers	Quadrants for gun carriages
Depth charge throwers	Rock drills
Friction spiders	Shell lathes
Gauges for 18lb and 4.3″ shells	Simple plummets
Greasers for caterpillar tractors	Spur wheels and pinions
High explosive shells	Storage tanks
Holman-Farrar hand air pumps	Tank dust collars
Howitzer bed pegs	Thrust rods and sockets

*Carry on! The patriotic image from the Holman book
'What we did during the Great War'.*

Trewin who was killed in Flanders after serving in Italy and Lieutenant Cyril Harvey, an ex-Holman apprentice in the R.A.F. posted missing.

By early in 1915 there was a very serious shortage of manpower throughout the Cornish industries. Old hands due to retire stayed on and others came back. Jebus Bickle, the Cornish engineer, happily came out of retirement to work in the Holman drawing office. James Miners Holman admitted later that, 'the contribution to the Empires fighting forces has greatly depleted our workshops of men. Many who formerly stood at the lathes and benches are today in the fighting 'Zones' on land or sea in the great conflict.'

Conscription further depleted manpower and the only solution was the wholesale recruitment of girls and young women. This revived a tradition of women workers in Cornish industry that had ended when the last 'bal maidens' left East Pool a few years before. Ben Cock at No. 3 Rock Drill Works rapidly found himself managing over 200, which perhaps accounts for the look of slightly sardonic resignation on his face when photographed amid his young ladies (see page 99). Their immediate foreman was ex-Tuckingmill Foundry apprentice James Parnall, who later belonged to the

Howitzer gun beds in No.1 Works awaiting packing.

Staff at the No. 3 Works in 1915 demonstrating some of their products.

BACK ROW: *Jack Trevaskis, Edwards, Moyle, Cowie, J. Major, S. Holman, S. Tregear*

MIDDLE: *Dick Nankivell, L. Penberthy, W. Johns, F. Godolphin, F. Uren, W. Jackson, ?, O. Rule, S. Williams*

FRONT ROW: *Curnow, H. Taylor, ?, H. Willoughby, Alf Phillips, Ben Cock, Arthur Thomas, George Pengelly, W. Rhodda, J. Oliver, J. Blewett*

for the manager of Boiler Works soon found himself with the Royal Engineers in Flanders where he served until gassed in 1917. Albert Bray, who began as a junior clerk in the head office, joined the Duke of Cornwall's Light Infantry and swapped cold Camborne for the heat of India. James Menhennet, the son of a manager in the Mysore Goldfields, had become an apprentice in February 1911. He had graduated to the drawing office when he volunteered for the Royal Navy, and served on H.M.S. *Lion* as an Engine Room Artificer during the Battle of Jutland in May 1916. Thomas Garfield Pidwell, who lived opposite the works, used to deputise for his older brothers who worked in the smith's shop, until he became an apprentice in August 1912. After working a hammer and driving the overhead electric crane, he volunteered in 1916, hoping to realise his great ambition of becoming a Guardsman. Instead, he served with the Devons until severely injured at the Battle of Cambrai in September 1918.

Many of course never returned to their workbenches. In one edition alone of *The Cornishman* in August 1918, there were obituaries for two Holmans men; W. C.

Managing such a complex and often fraught industrial operation rested almost entirely upon James Miners Holman and his nephew Leonard, who, with deep reluctance, had been forced to stay behind while his brothers and cousins served in France. 'Mr Leonard' always regarded his part in the war as 'inglorious', yet he was absolutely vital to what had become a busy armaments works. However, this did not save him from having a 'white feather' thrust into his hands, a cruel and ignorant gesture typical of the times, when people still failed to appreciate that allowing skilled men and managers to enlist might cripple the very industries needed to prosecute the war.

Many had already volunteered or gone with the Territorials and 126 Holmans men would serve in the forces, five of them being commissioned. Mr Kenneth joined in the Royal Artillery where he served for the duration. Private Johns from Tehidy Road was among the anti-aircraft gunners who shot down 'L 15', the first Zeppelin to fall on Britain. The young man who daily drove the pony trap

Young woman operating a lathe in 1915.

Another group of women workers at Holmans in 1914.
These seem to be having more fun than in the previous image.

France. He spent the remainder of the war flying the famous Sopwith Camels; on ground attack or often low-level missions when the circling SE5s and Bristol F2b 'Brisfits' drove the Huns down to the waiting Sopwith Camels. Treve never lost his love of flying and years later, even when travelling the world as a successful businessman, he invariably managed to visit the flight deck of the airliner.

Meanwhile Holmans, along with every other Cornish engineering works, had been requisitioned by the government and production lines were reorganised and retooled for war. Ordinary work, except for rock drills for the South African gold mines, virtually ceased. Where mining machinery had once occupied the benches and shop floors, there now appeared a variety of munitions; bomb gauges, powder moulds, tripods for Vickers machine guns, howitzer base plates, paraffin engines and tracks for the amazing new 'tanks'. One closely overseen section manufactured H.E. Shells and No. 2 Works was filled with trawler pattern diesel engines to equip submarine chasers for the French navy. Presses and machinery were also supplied to Bickford-Smith and the National Explosives Company, both heavily engaged in making munitions.

1914 through early 1915 a good deal of reorganisation was carried out on the Wesley Street works, in response to the requirements of war. This saw the transformation of the works by the demolition of the old small workshops surrounding the yard on the west side, next to the Chapel, and the construction of three large bays under a single roof. This would become the Heavy Machine Shop. The work was designed and supervised by Nicky Curry, using the structural engineers Gardiners and glazing specialists Helliwells.

Despite James Miners Holman wanting his nephew to remain at home and run the increasingly busy foundry, Treve was all the more determined by his brother's death to join the army. He enlisted in the Royal Field Artillery but he was not destined to go to France, where his brother Kenneth and cousin Percy were in the artillery. Instead he was posted to Mesopotamia. However, he still longed to fly and later transferred to the Honourable Artillery Company, then a training cadre for officers. By early 1918 he was in the Royal Flying Corps and got his wings in June, two months after it became the R.A.F. He was disappointed not to have his wings from the old R.F.C., but was happily transferred to No. 54 Squadron in

Women workers at Holmans in 1914.

there are cases of brutality but there are rotters in every army.

Just eight days later Jack Holman was shot through the head by a sniper while trying to rescue a brother officer in the trenches near Armentiers. He endured the long trip back to the field hospital by truck and was taken by train to the main military hospital at Boulogne, meeting an unnamed old school friend on the way. He never recovered and died of his wounds on 30th October 1914 and was buried in the Boulogne Eastern Cemetery. One particularly tragic episode was that his sister went across to nurse him, or at least she believed so. However, his head was heavily swathed in bandages and it was not discovered until after his death that there had been a dreadful mistake, for the wounded man was not Jack but another officer named Holman. Through late

Holman stand at the Anglo-American Exposition, July 1914.

Dragoon Guards were soon sent to France, where the regiment was credited with the first action by the British army in the war. They saw considerable fighting during the first few months and, like many young Cornish soldiers, Jack wrote home and on 21st September 1914 the *Cornish Post & Mining News* published 'another cheery letter from Lieutenant Holman', written while his regiment, dismounted to serve as infantry, were resting

> after being heavily engaged yesterday. We relieved the infantry and were under heavy fire for some time. We were mentioned in despatches and our major has been recommended for his VC. He is an absolutely topping leader without any fear. For the last two or three days I have not felt very well as I caught a chill from getting soaking wet and then sleeping in my wet clothes. By the way the report of the Germans treatment of our wounded and prisoners are exaggerated colossally. In fact they fight fairly though

In the spring of 1914 the situation was worsened by the sudden collapse of the over inflated tin market, which had been heralded by the collapse of the ephemeral empire of Cornish Consolidated Tin Mines Limited and almost all of their highly expensive reworking of old Cornish mines in 1912. A lot of Holman machinery soon went up for auction, including the big air compressor at Botallack and the 400 h.p. cross compound winder, built for the 1909 exhibition and installed with six heads of pneumatic stamps at Great Dowgas Mine. At South Phoenix Mines the 80-inch pumping engine, christened so hopefully by the Prince of Wales, went out of steam for good while in June 1914, Woolf"'s 80-inch pumping engine ceased work at Condurrow (this engine had worked at two other mines and all three engine houses still stand). Few realised she would end up as assorted scrap iron in the foundry furnaces at Holmans.

The unhappy state of the Cornish mines was temporarily forgotten when war came a few months later in August 1914. Jack Holman and the 4th Royal Irish

General view of the No.1 Works Erecting Shop.

LIST No. S.4./25

THE "HOLMAN"
"TWIN-GRIP" No. 10
DRILL SHARPENER

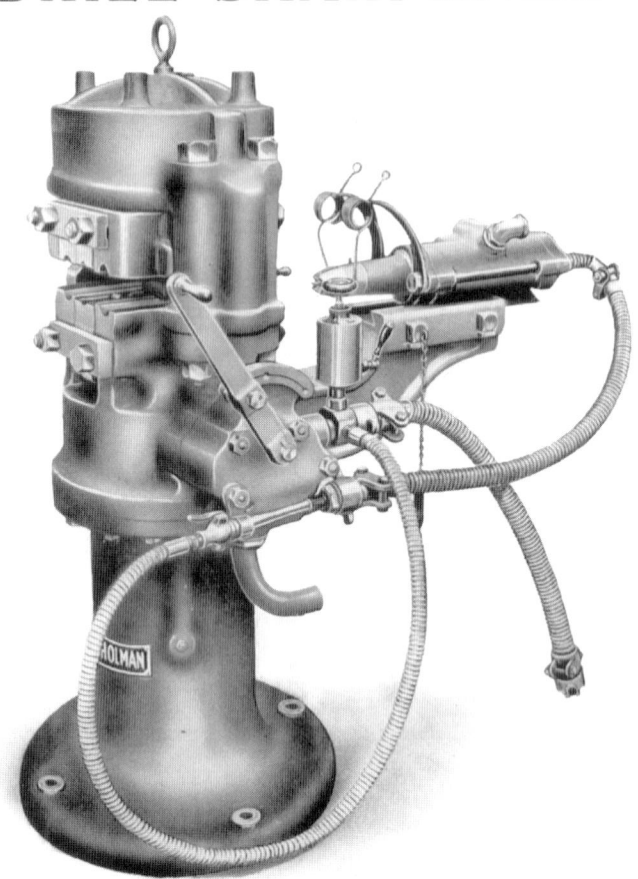

Holman BROS LTD.

CAMBORNE   **ENGLAND**

London Office: BROAD STREET HOUSE, E.C.2

For rock drills there was the rock drill sharpener, the essential accessory. These were manufactured and exported in their thousands.

when the pilot mistook his waving to go higher as a signal to descend. He did fly again during the hot summer days of June 1914 when the young Irish earl Lord Carberry, the 'Flying Earl', brought his 80 HP Morane to Sinns Barton Farm near Redruth. Kenneth, Treve and Leonard willingly paid their five guineas for a flight, as did Miss Dorothy, Leonard's sister, but only one gentleman from Penzance was prepared to pay the thirty guineas to loop the loop.

A souvenir card of Henri Salmet's visit to the West Country in 1912.

James Miners Holman himself, as chairman of East Pool, was attempting to settle an acrimonious dispute between the committee and the miners over reform of working hours and payment of wages. Holman had known for a long time that such a big and relatively modern mine could not survive under the old 'cost book' system and that it would have to be replaced by a limited company. Accordingly he had invited W. C. Moreing of Bewick & Moreing, a large mining consultancy, to join the board and help carry through the change over. Their partnership saved East Pool and James Holman's daughter Dorothy actually married Captain Algernon Moreing, who became a Member of Parliament for Buckrose in 1918-22 and for the Mining Division (Camborne) in 1922-23 and 1924-29.

while down at the Cornwall Boiler Works, a new oil fired furnace capable of bringing a two inch bar to white heat in a few minutes had been installed. There was a battery of big hammers, including a 12 cwt die headed which once tripped 'One, two, three, the result was a flattened pad which would later form part of a rock drill'. Parted off under the cutting hammer, the steel was de-frazed by a press 'leaving in the tongs a red hot but finished drop forging'.

John Warren, the manager, was extremely proud of his new machinery which cut working time from hours to minutes and was capable of producing 200 forgings a day, making Boiler Works one of the most modern works outside the Midlands. He was a stern and exacting engineer who was driven back and forth to work each day in a pony trap by a young lad who would himself join the works. Fifty years later he recalled that a gentleman who wanted a completed boiler delivered to Hayle, on being asked how he wished to pay, opened up an 'old green bag' full of shining gold sovereigns. Another boiler contract for Hayle provoked the anticipated storm from John Warren when he discovered that the company concerned had shut without cancelling the order. More bizarre was the story of the heavy boiler which was sent to St Just hauled by a team of thirty horses. Left to its own devices overnight the boiler disappeared and was never seen again. Rather optimistically it was presumed that it had rolled over the cliff.

The next generation of Holman boys, like their fathers before them, had joined the family tradition that extended from boardroom to shop floor. That is with the exception of John's son Jack, a splendid and happy young man who, after leaving Blundell's College at Tiverton, had chosen a military career, and gone off to Sandhurst. In September 1913 he was gazetted a Lieutenant in the 4th Royal Irish Dragoon Guards and was posted to Dublin, enjoying the polo playing life of a young officer.

Treve Holman was an energetic and athletic twenty year old who, like many young men with a mechanical bent, was captivated by the new art of flying. Having failed to get airborne because of rough weather at Hendon in 1911, he went aloft with the French aviator Henri Salmet (also known as the 'Daily Mail Aviator), who thrilled crowds by flying from the Eastern Green beach at Penzance in May 1914. Circling St. Michael's Mount at 1,400 feet was also Treve's way of celebrating that he had just been elected to the Board of Holmans.

Treve Holman eventually managed to go flying at Hendon but was disappointed

Female worker during the Great War.

Chapter Five

The Great War

The splendid response of the many hundreds of girls who in our rock drill department alone took the place of over two hundred skilled men, was a factor of considerable importance.
From: *What we did during the Great War.*

The momentous year of 1914 began splendidly for Holman Brothers Ltd with a full order book, especially for the Rand. Other machinery and drills were also due to be shipped out to India, China and Russia. One hundred and eight South African mines and 104 Australian mines used air, steam or electric hoists which could be adapted for pumping as well as winding. Holmans new catalogue reflected their energy and adaptability as, following complaints from British Consuls that only the German companies produced publicity in foreign languages, they published another catalogue in Spanish, fully illustrated and on fine art paper. Two 'Holman Scholarships' worth £100 per year were created for boys attending the Camborne elementary schools, the company providing an ordinary apprentice wage for five years and the fees for attending the mining school at Camborne. The company also found the opportunity to exhibit their latest rock drills at the Anglo-American Exposition held in London during July 1914, where the model of William Murdoch's tiny steam 'Flyer' was also exhibited.

Of enormous importance to all the manufacturers of rock drills was that 1914 saw the end of the patent on the American Leyner rock drill, which had gone a long way towards solving the underground dust problem.

The expansion of No. 3 Works saw rock drill production being almost doubled,

His father William Cox was not only chief draughtsman for Holmans but also an exceptional artist himself, who had many friends among the Newlyn School of painters. Indeed, he named his son after the celebrated painter Norman Garstin. The young Garstin Cox had studied at the Camborne School of Arts and later with John Noble Barber of the St Ives School. His magnificent Cornish landscapes and seascapes were shown at the Royal Academy and when they were portrayed on the Holman calendars, were pinned on walls by 'Cousin Jacks' in mining camps all over the world.

There was also a visit from Monsieur Kinsman, a French safety fuse maker who was somewhat overwhelmed as he found that

> by the means of a splendid array of machine tools, each of which cost some hundreds of pounds.... which enables them to turn out from 80 to 100 complete drills and spare parts per week. Many of their machines are automatic, men and machines work rapidly in ideal conditions as regards light - air and space. A part can be made to a thousandth of an inch. At their No. 1 Foundry I learned that having settled in their new offices they are gradually demolishing all the factory in the square leading west to Centenary Chapel, and that this will be covered with new buildings, all the present yard space being roofed over, so that in making pneumatic stamps, winches, air compressors, and other heavy machinery, the conditions will be as modern and efficient as at their rock drill works.

It was a splendid accolade for Holman Brothers Limited, which never seemed more prosperous or bustling, while the family's ordinary business was supplemented by an extraordinary share portfolio. Besides the Malayan tin fields, Brazilian gold mines and Cornish utilities and banks, it also embraced such novelties as the Red Sea Phosphate Company and the Sao Paulo Tramway, Light & Power Company. Yet within a year much was to change and a new era was about to begin for the company.

industry. By January 1913, Wheal Grenville was building the foundations for its own Holman compressor, driven by a 230 HP motor built by the Lancaster Dynamo Company. Both compressors were to meet up in their later careers.

Holmans also managed one old time job when the bob of the stamps engine at Wheal Kitty near St Agnes suddenly broke on Friday 13th. May 1913. A new one was ordered that same evening and left the yard of No. 1 Works at 11 am just eight days later. At 10.30 am on 25th June the bob was raised into position, which J. H. Collins, the doyen of Cornish mining, considered 'a fine piece of repairing work, which I think holds the record for work of this class... and a great credit is due to Messrs Holman'.

View of the No.3 Works with its many varied belt-driven machines.

The company calendar for 1913 celebrated their achievements world wide with the painting 'Evening in a Cornish valley' by Garstin Cox, Ben Cock's nephew, who at only twenty-one was maturing into a superb artist. He was a rare being, a native Cornish artist, whose roots were in the 'Black Country' around Camborne.

There were still plenty of orders for big steam winders like that supplied to the Pachuca Mines in Mexico, home of generations of Cornish miners. Captain Josiah Paull of South Crofty, a mere fortnight after ordering two self dumping skips, accepted Holmans tender for a paired 22-inch cylinder steam hoist with a 48-inch stroke, complete with two 8 feet by 3 feet winding drums, drop valve gear, cage indicator and foot operated brakes. He requested that 'Delivery to be made to New Cook's Shaft and the services of a skilled erector to be provided ... We are prepared to increase the time required for delivery to five months providing you will accept this date under a penalty clause of 1% per week.' The new hoist took slightly longer, and not until 1st April 1914 could Captain Paull record that 'the erection of the new winding engine is well in hand and we hope to resume sinking operations by May 1st.' A week later, the new winder was baling out the shaft with water skips ready for the work.

Holman had also recently supplied East Pool with a new air compressor capable of draughting 1,200 c.f.m. and driven by a 215 HP Westinghouse electric motor. This was fed by the grid of the Cornish Electric Power Company whose new power station at Hayle had resulted in an 'electric boom' amongst Cornish

A small boiler just completed at the Boiler Works.

Holman stretcher bar hoist in use at Dolcoath Mine..

settled into an increasingly eccentric old age with her backward son and the estate eventually fell into Chancery. James Holman restored the house to its Georgian elegance and where the grounds faced Tehidy Road, an imposing stone wall soon replaced a crumbling hedge with an impressive new house and iron gates made at the No. 1 Works.

James Holman's continuing prosperity reflected that of the works which already had a wide reputation of improvement and innovation. Story has it that while homeward bound from South Africa he was joined in the steamer's saloon by some mining engineers and Cornish 'Cap'ns' returning from Australia. They discussed improvements to machinery and he was asked, 'Why don't you build a hoist light enough to clamp on a simple drill column and powerful enough to hoist 500 to 1,000 pound'. The result was the Holman 'stretcher bar hoist' which weighed just a little more than the 3¼" drifter, was driven by either single or double air cylinders and could be erected in very cramped places. The new hoist was soon widely used, especially in Australia.

the U.S.A. There, he spent some years as an improver in a railway locomotive shop at Duluth, Minnesota before returning home to join Holmans as a lathe operator and then going over to No. 3.

In 1911 James Holman acquired a more palatial residence, Rosewarne House, which was built in 1773 by the Harris family of Camborne. Their eldest son William was a contemporary of Richard Trevithick, whose famous Road Locomotive ran past the gates of Rosewarne on Christmas Eve 1901. Eventually Rosewarne came to an only daughter who, having made an unhappy marriage,

Taken in the 1930s this aerial photo shows how close Rosewarne House was to the Holman Works. The house is near the top centre (1), No.3 Works bottom left (2) and No.1 Works bottom centre and right (3).

It was in the early 1900s that the first portable compressors were produced in Camborne. This one is in the erecting shop.

James McCulloch, after sixty years. The Holmans do not appear to have attended the three-day auction, though they must have derived a certain satisfaction when his Little Hercules drills were sold off cheaply in batches. Many of the skilled men and apprentices, like those who came from Harvey's, were happily re-employed by Holmans. William Yeo started at the age of thirteen in the moulding shop at Tuckingmill, then spent six years at James Hosking's Foundry which was demolished to make for Holmans new rock drill works and, after working at Roseworthy Hammer Mills, joined No. 3 in 1911. Charles Blewett was originally a Harvey apprentice who graduated to Tuckingmill before joining Holmans No. 3 Works in 1911. After leaving Roskear School, Jack Powning served at Tuckingmill until joining No. 3 where his first job was assembling and testing new cradle hammer drills for America, and he would spend the next fifty years with Holmans. Another Tuckingmill apprentice was Charles Trezona Phillips who, once time served, migrated to

a record for their Imperial drill for single handed driving a level in New Primrose and this was not the last of the rivalry between the two companies.

The gold mines of the Witwatersrand would remain one of Holmans best customers for decades, although they continued to export mining machinery around the world. In May 1905 their 3½" inch drills sank the main shaft of the Murchison Mine in Australia 81 feet in a month through very hard diorite, despite stoppages for pumping and timbering. Holman pneumatic stamps were installed, at £1,000 less than a Californian battery, at the Mountain Queen Mine in Western Australia as part of a massive crushing complex with a 150 ton ore bin, sieve loaders, and line shafting mounted on tall concrete pillars; the whole operation being housed in a huge galvanised shed.

The continuing success and increased expansion of the Camborne business had led the Holman Brothers to convert to a private limited company in 1906, with themselves as sole directors. Unfortunately John, the senior director, was already very seriously ill and in the following year James's son John Leonard Holman, who had just finished his apprenticeship under the redoubtable Nicholas Curry, joined the board. 'Mr Leonard' was soon dispatched on a worldwide business trip and in 1910 alone he visited Mexico, the U.S.A. and Canada.

The new pneumatic stamps, 1910.

However, he was home in September to marry Gladys Fredericka Thomas, daughter of C. Fred Thomas and granddaughter of the great Captain Josiah Thomas of Dolcoath and whose mother was from the Bartle family of iron founders.

October 1910 saw the closure of the Tuckingmill Foundry, once the lair of

large pile of ingots of Cornish tin from the Williams Harvey smelter. Sadly the monument has not survived for posterity and neither did the beautiful bowl made by the Newlyn Art Classes from silver raised from the Perran Mine during 1908. James Miners Holman who been travelling with his wife in Egypt for the past two months arrived home in time to attend the 'Cornish Day' at White City which, held on the lawn in bright sunshine and hosted by Lord and Lady Falmouth, was a splendidly Cornish occasion.

The Great Stope Drill Contest was finally won by Holmans, whose 2⅛", after a rigorous 12 months of testing, drilled 12,779 inches at a cost of barely 10 pence a foot, then half the cost of black labour. The cheque was presented by the Minister of Mines for the Transvaal in September 1910, just after six Holman 3¼" inch machines excelled their past records by sinking a 20 feet by 8 feet incline shaft 279 feet in a month at Van Ryn Deep.

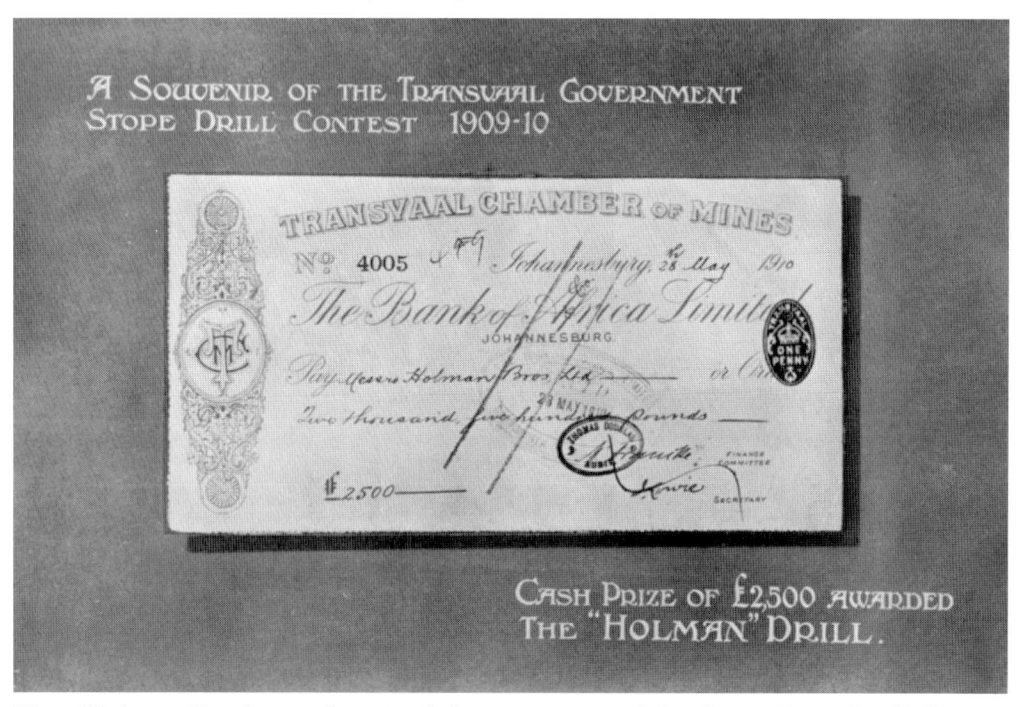

How Holman Brothers advertised their winning of the Great Stope Drill Contest in one of their catalogues.

The two competition drills were shipped home and displayed at the Polytechnic Society Exhibition where it was noted that 'the Rand dust is still upon them'. Holman gained a Diploma of Honour but back on the Rand, Climax soon claimed

was to 'hold the chuck over a fire of old candle ends and dynamite boxes'; in desperation one shift boss used a shaving of dynamite to blow the drill steel out of the chuck.

The Great Stope Drill Contest began in May 1909 with an added bonus of substantial prizes for the those miners manning the first, second and third machines. The trials, at Transvaal University and underground in Feirrera Deep, eliminated nine of the nineteen competitors but low air pressure, the curse of the Rand mines, meant the 300 shifts were reduced to 217 and the competition was extended by a further six months. The struggle waged back and forth in the gold stopes of Crown Deep, Village Deep and Robinson Deep, with the laurels for the best machine being reluctantly relinquished each time some other machine bored deeper and faster.

The marathon competition was keenly followed back in Cornwall, though the partisans of both Climax and Holmans had to share the 'Cornish Pavilion' at the Imperial Exhibition held at White City during the summer of 1909. This was a glittering showcase for Cornish industry with the Prince of Wales as patron and well supported by the mining gentry and mineral landlords. All the big firms displayed their products. Holman showed a variety of their drills, air cushion stamps and the popular stretcher bar hoists, which were being used in 100 mines in Africa and Australia, Climax had their Imperial rock drill and Tuckingmill Foundry, the 'Little Hercules'. National Explosives and Bickford-Smith mounted large displays while many working Cornish mines sent minerals and ores. R. Arthur Thomas eclipsed them all with a model of Carn Brea monument cast in solid gold to represent the total value of tin and copper raised during the long career of Dolcoath Mine. Flanking this amazing work of art was a big ebonised case full of Dolcoath minerals and a

> **ROCK DRILLING ON THE RAND.**
>
> THE "HOLMAN" STOPING DRILLS WIN IN THE TRANSVAAL GOVERNMENT CONTEST
>
> Inaugurated by the Chamber of Mines, 1909-10.
>
> THE FIRST PRIZE OF £4,000 AND SECOND PRIZE OF £1,000 DIVIDED BETWEEN THE 'HOLMAN' AND THE 'SISKOL' MACHINES.

Another drilling team at Brakpan Mine.

The South African Chamber of Mines joined the fray when in February 1908 they offered a prize of £2,500 for the best drill that was sufficiently light and small to work the narrow gold stopes. This was doubled by the Transvaal Government and a formal competition was organised under strict rules whereby all drills would undergo surface trials at the Transvaal University and then complete 300 shifts, 43 each in a working gold stope. Once the footage and individual costings were established, it was believed that all but the best machine would be eliminated.

Even as the competition was being organised, Holmans claimed another record when their 3¼" drills drove a level 226 feet in Cinderella Deep during March 1908. This was quickly followed by another at Brakpan Mine when their big No. 1 Incline shaft was sunk 223 feet by Holman drills during June 1908. Such pounding inevitably took its toll of any rock drill, despite the maker's claims. One Randite considered the old Holman tappet machine one 'of the worst he had ever used, especially its patent tapered chuck'. Unless lined with an old tin can it usually jammed solid at the end of a shift and the only way to free the drill steel

opposed to the enormous reciprocating drifters which, even after twenty years of development, remained great consumers of air, time and manpower. There now ensued almost a repeat of the frantic rivalry between rock drill companies, which had happened in Cornwall 25 years before during the early days of boring machines. To bring order into the chaos of competition, various trials were arranged. The *South African Mining Journal* sponsored a contest in the winter of 1907 but in a startling result the Gordan drill, made by Ingersoll Sergeant, beat all comers though the American machine did not fulfil expectations in practical use underground.

All previous Price Lists of Rock Drills and Accessories are hereby cancelled.

ADDRESSES -

CAMBORNE:
Telegraphic and Cable Address: "AIRDRILL, CAMBORNE."
National Telephone, No. 7, CAMBORNE.

LONDON -
BROAD STREET HOUSE, E.C.
Telegraphic and Cable Address: "AIRDRILL, LONDON." Telephone, 9270 LONDON WALL.

TRANSVAAL -
Box 619, Corner House, Commissioner Street, Johannesburg.

AUSTRALIA -
AUSTRALIAN METAL CO., LTD., 113, William Street, Melbourne. Box. 221.

MEXICO -
MESSRS. LUIS MONROY DURAN & Co., Apartado 1607, Calle de gante, No. 1.

RHODESIA -
MESSRS. S. SYKES & Co., Ltd., Box 230, Willoughby's Buildings, Bulawayo.

CANADA -
MESSRS. MUSSENS, Ltd., 318, St. James Street, P.O. Box 2519, Montreal.

CODES -
A.B.C.; BEDFORD McNEIL; and MOREING AND NEAL.

Page from the Holman Brothers catalogue of 1910 showing the offices already established across the mining world.